Workbook 4

Hydraulic Fluids Conditioning

Dr. Medhat Kamel Bahr Khalil, Ph.D, CFPHS, CFPAI.
Director of Professional Education and Research Development,
Applied Technology Center, Milwaukee School of Engineering,
Milwaukee, WI, USA.

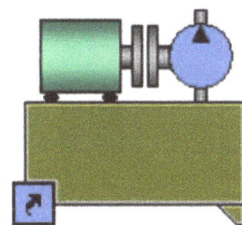

CompuDraulic LLC
www.CompuDraulic.com

CompuDraulic LLC

Workbook 4

Hydraulic Fluids Conditioning

ISBN: 978-0-9977816-7-0

Printed in the United States of America
First Published by Sept. 2022
Revised by Feb. 2024

All rights reserved for CompuDraulic LLC.
3850 Scenic Way, Franksville, WI, 53126 USA.
www.compudraulic.com

Disclaimer

It is always advisable to review the relevant standards and the recommendations from the system manufacturer. However, the content of this book provides guidelines based on the author's experience.

Any portion of information presented in this book could be not applicable for some applications due to various reasons. Since errors can occur in circuits, tables, and text, the publisher assumes no liability for the safe and/or satisfactory operation of any system designed based on the information in this book.

The publisher does not endorse or recommend any brand name product by including such brand name products in this book. Conversely the publisher does not disapprove any brand name product by not including such brand name in this book. The publisher obtained data from catalogs, literatures, and material from hydraulic components and systems manufacturers based on their permissions. The publisher welcomes additional data from other sources for future editions.

Workbook 4
Hydraulic Fluids Conditioning
Table of Contents

PREFACE

This Workbook is a complementary part to the textbook of the same title. This book is used as a workbook for students to take notes during the course delivery. It contains colored printout of the PowerPoint slides that are designed to present the course. Each chapter is followed by a number of review questions and assignments for homework.

Dr. Medhat Kamel Bahr Khalil

<table>
<tr><td colspan="2" style="background:orange"> Chapter 1
Introduction </td></tr>
</table>

Objectives:

In addition to hydraulic fluids contamination control, conditioning of hydraulic fluids is an essential process for a hydraulic system to perform properly and reliably. This chapter introduces the scope of hydraulic fluids conditioning.

Brief Contents:

1.1- Scope of Hydraulic Fluids Conditioning

1.2- Scope of the Textbook

0

0

1.1- Scope of Hydraulic Fluids Conditioning

❖ Hydraulic Fluids Conditioning: →

▪ Fluid Analysis.

▪ Contamination Control:
  ○ Removal of air (reservoir design, and foam suppressing additives).
  ○ Removal of water (absorptive filters, vacuum dehydrators, and centrifuges).
  ○ Removal of chemical (acids, sludges, varnish, oxidation products, etc.).

(Already discussed in Volume 3)

1

1

1.2- Scope of Textbook

❖ Hydraulic Fluids Conditioning (in this textbook):→

- Hosting (Hydraulic Reservoirs).
- Sealing (Hydraulic Sealing Elements).
- Transmission (Hydraulic Transmission Lines).
- Temperature Control (Heat Exchangers).
- Removal of Particulate Contaminants and Silt (Filters).

2

2

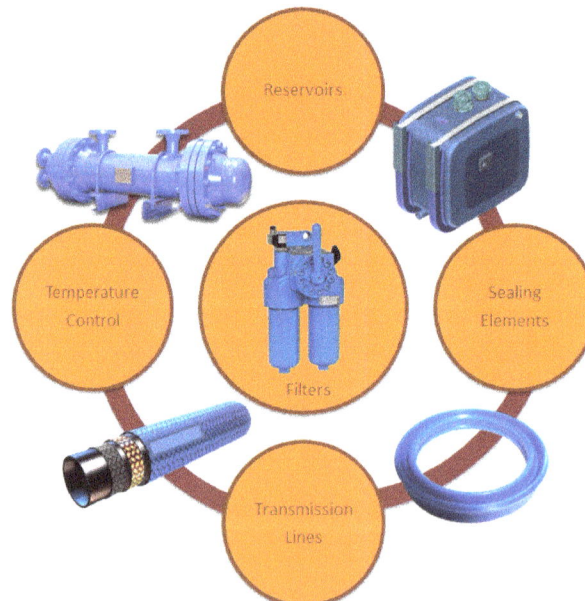

Fig. 1.1 – Hydraulic Fluids Conditioning

3

3

Chapter 1 Reviews

None

Chapter 1 Assignment

Student Name: -- Student ID: ------------------

Date: --- Score: ------------------------

Assignment:

In your understanding, what are the actions of hydraulic fluids conditioning.

Hydraulic Reservoirs

Objectives:

Unless a hydraulic reservoir is designed, constructed, installed and maintained properly, the reliability of the entire system will be adversely affected. This chapter presents an overview of the different types of hydraulic reservoirs for industrial and mobile machines. This chapter also provides guidelines for design and sizing of hydraulic reservoirs.

Brief Contents:

2.1- Contribution of Hydraulic Reservoirs

2.2- Configurations of Hydraulic Reservoirs

2.3- Construction of Hydraulic Reservoirs

2.4- Design of Hydraulic Reservoirs

2.5- Hydraulic Reservoir Design Case Study

0

0

The following topics are discussed in Chapter 9 in Volume 5 "Safety and Maintenance"

- Replacement
- Scheduling
- Installation and Maintenance

The following topics are discussed in Chapter 9 in Volume 6 "Troubleshooting and Failure Analysis"

- Inspection
- Troubleshooting
- Failure Analysis

1

1

2.1- Contribution of Hydraulic Reservoirs

❖ **Terminology:** Reservoir OR Tank

❖ **Main Contribution (1): Hosting**

Video 236 (2 min)

Video 639 (1.5 min)

❖ **Auxiliary Duties:**
- Remove Contaminants (2).
- Remove Air (3)
- Remove Heat (4)
- Equipment Base (5)

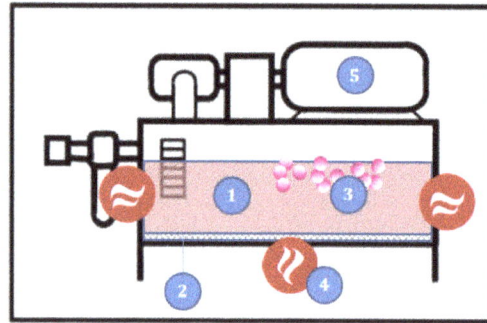

Fig. 2.1 – Contribution of Hydraulic Reservoirs

2

2

2.2- Configurations of Hydraulic Reservoirs

Pressure	Shape	Volume	Application	Function	Material	Size
Open	Parallelogram	Fixed Volume	Standard (Industrial)	Main Reservoir	Traditional (Steel)	Compact & Small
Closed	Cylindrical (Custom)	Variable Volume	Custom (Mobile)	Add-On Reservoir	Hybrid	Medium & Large

Table 2.2 – Configurations of Hydraulic Reservoirs

2.2.1- Open versus Closed Hydraulic Reservoirs

- **Open:** applications (conventional industrial and less contaminated).
- **Closed:** applications (special mobile and highly contaminated).
- Back pressure → help avoid cavitation.

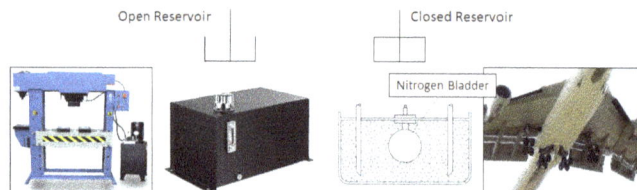

Fig. 2.2 – Open versus Closed Hydraulic Reservoirs

3

3

2.2.2- Parallelogram-Shaped versus Rounded Reservoirs

❖ **Cleanliness:**

▪ Standard Reservoirs:

▪ flat surface and sharp corners → contaminants settlement .

▪ Cylindrical Reservoirs:

▪ Suction connection to the apex of the inverted cone.

▪ Return flow inlet is tangential to the cylindrical wall.

▪ → Flow returned dynamically

▪ → Contaminants are easily trapped by filter.

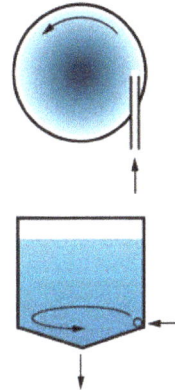

Fig. 2.3 – Cylindrical Hydraulic Reservoirs
(www.powermotiontech.com)

4

❖ **Cost Effectiveness:**

▪ Standard Reservoirs:

o Hydrostatic pressure on sharp corners & flat surfaces →

o Side plates deflection outward → welded supporting ribs →

➤ Extra cost.

➤ Inside Ribs → contamination collected in corners

➤ Outside Ribs → additional floor space.

▪ Cylindrical Reservoirs:

o Better pressure distribution → less wall thickness → less cost.

5

2.2.3- Fixed versus Variable Volume Reservoirs

❖ **Concept:**
▪ **Fixed Volume Reservoirs (FVR):**
o Volume of Oil: fill components + operate the system + extra for cooling.
o Relatively large and heavy.

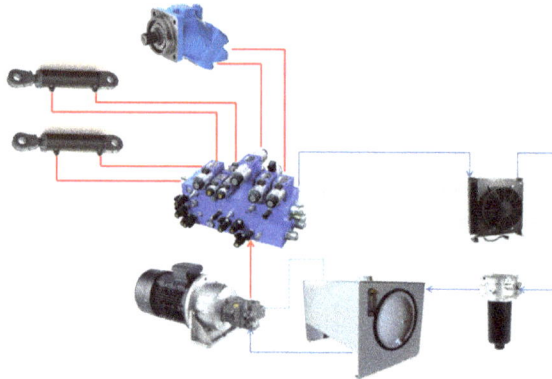

Fig. 2.4 – Typical Simplified Hydraulic Circuit with Conventional Fixed Volume Reservoir (Courtesy of Smart Reservoirs)   6

6

▪ **Variable Volume Reservoir (VVR):**
o Volume of Oil: thermal expansion + differential cylinders + accumulator.
o Compact and Installed near the pump
▪ Weight reduction can be up to 100:1 compared to classic solutions.

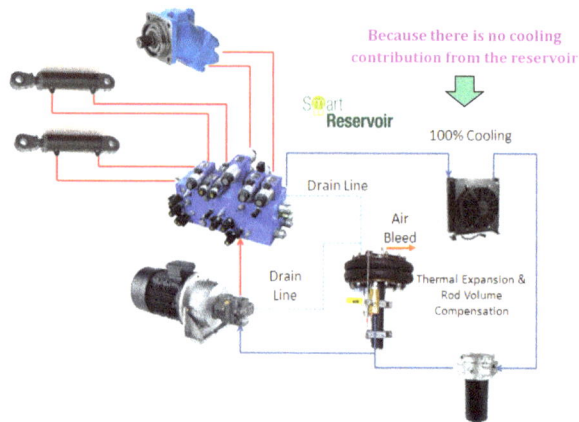

Fig. 2.5 – Typical Simplified Hydraulic Circuit with Variable Volume Reservoir (Courtesy of Smart Reservoirs)   7

7

❖ **Applications:**

▪ Used for both open and closed circuits.

▪ Mobile, military, aerospace, marine and offshore fluid power systems.

  ❖ **Closed Circuits (Hydrostatic Drives):**
   ▪ Save the cost of boosting pump.

    📹 Video 397 (0.5 min)   📹 Video 398 (0.5 min)

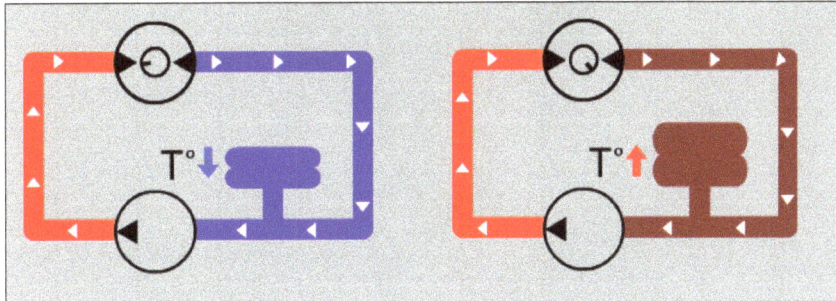

Fig. 2.6 - Operating Principle of Variable Volume Reservoirs in a Closed Hydraulic Circuit (Courtesy of Smart Reservoirs)

8

8

❖ **Open Circuits:**

    📹 Video 395 (0.5 min)   📹 Video 396 (1.0 min)

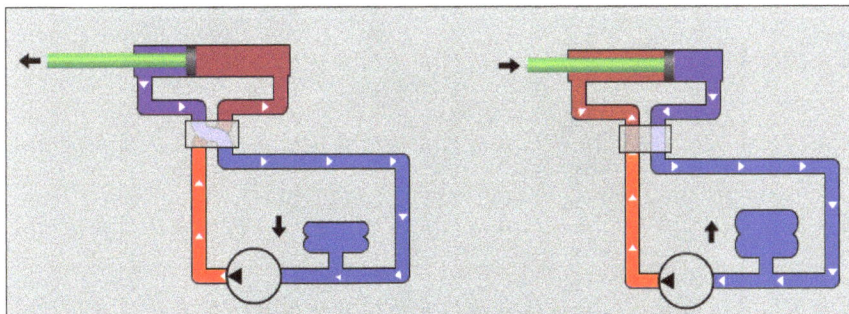

Fig. 2.7 - Operating Principle of Variable Volume Reservoirs in an Open Hydraulic Circuit (Courtesy of Smart Reservoirs)

9

9

❖ **Construction (1):**
- Slightly pressurized bellows to 0.1 to 0.6 bar (1 to 9 psi).
- Wall elasticity, not gas pressurized.

Fig. 2.8 - Construction of Variable Volume Reservoir
(Courtesy of Smart Reservoirs)

10

10

❖ **Construction (2):**
A. Anodized aluminum cover.
B. Built-in low-level switch.
C. Air bleed valve.
D. Filtration/bleed return with isolation ball valve.

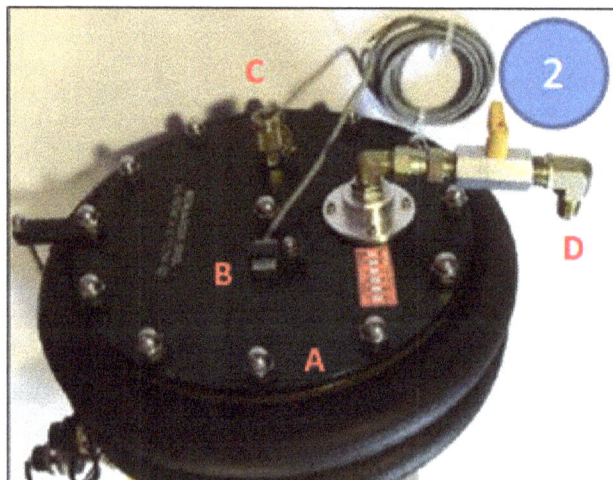

Fig. 2.8

11

11

❖ **Construction (3):**

A. Anodized aluminum base manifold.
B. Connection port.
C. VVR coupling fitting.
D. Air bleed valve.
E. Filling coupler.
F. Adjustable temperature switch.
G. Visual temperature indicator.
H. Pressure gauge.
I. Relief valve.

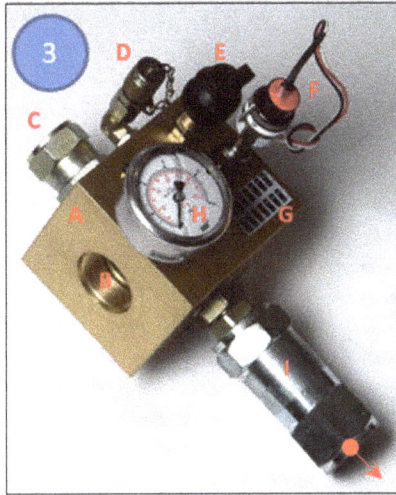

Fig. 2.8

12

12

❖ **Construction (4):**

Full assembly of VVR with the instrumentation manifold.

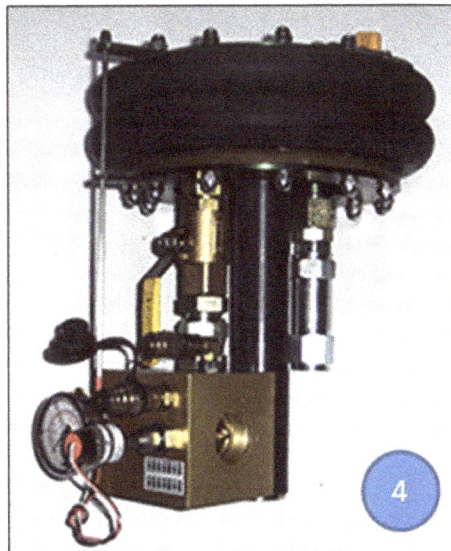

Fig. 2.8

13

13

❖ **Installation of VVR:**
- A hand pump is used to fill in the system with the required volume of oil.
- All drain lines are directed to the VVR via an in-line filter.
- Accumulated air → air bleed valve.

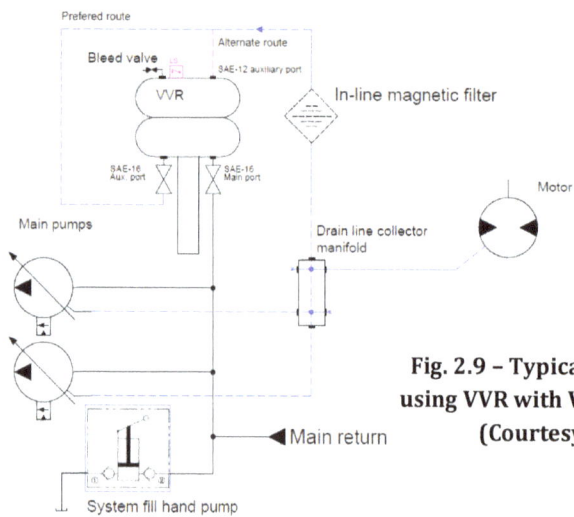

Prefered route

Alternate route

Bleed valve

SAE-12 auxiliary port

VVR

In-line magnetic filter

SAE-16 Aux. port

SAE-16 Main port

Motor

Main pumps

Drain line collector manifold

Fig. 2.9 – Typical Simplified Circuit Diagram using VVR with Variable Displacement Pumps (Courtesy of Smart Reservoirs)

Main return

System fill hand pump

14

14

❖ **Sizing of VVR:**

❖ FVR →
- Capacity ≈ (2-3) Qp.

❖ VVR →
- Capacity is independent of Qp.
- Capacity α
  - Differential cylinder.
  - Accumulators.
  - Thermal expansion.
- Multiple VVR can be connected in series or in parallel to increase volume.

Fig. 2.10 - Sizes of Variable Volume Reservoir (Courtesy of Smart Reservoirs) 15

15

❖ **VVR versus FVR:**

▪ **Reduced Weight and Space:**
○ Weight Ratio (100:1) and Space Ratio (20:1) → mobile applications.

▪ **Pump Supercharge:**
○ VVRs are slightly pressurized →
○ Improve pump inlet conditions regardless of altitude and orientation →
○ Applications (Aerospace and under water applications).

▪ **Pump Suction Line Size:**
○ VVRs are slightly pressurized → Avoid cavitation → **Ds** reduced by half.

▪ **Reduced Contaminant Ingestion:**
○ VVRs are isolated from harsh environments →
○ Less ingression of (solid particles, moisture, dust, etc.).

16

▪ **Lower Maintenance Cost:**
○ VVRs are isolated → filter replacement ↓ → maintenance cost ↓
○ VVRs are small size →
○ Easier leak detected → spillage and maintenance cost ↓

▪ **Environmental-Friendly:**
○ VVRs are small size → promotes use of expensive biodegradable fluids.

▪ **Fluid Compatibility:**
○ VVR is used with most mineral and biodegradable fluids and water glycol.

17

❖ **VVR versus Accumulators:**

▪ **Pressure Sensitivity:**

○ VVR reacts to fractional pressure variations.

○ Accumulators are not that sensitive.

▪ **Temperature Variation:**

○ VVR performance is not affected by temperature variations.

○ Accumulator's performance are affected by temperature changes.

▪ **Weight and Volume:**

○ For same fluid displacement, VVR weight is << accumulator

○ Accumulator (fluid volume + gas volume + heavy shell).

▪ **Simplicity:** The VVR is leak free, gas free and requires no maintenance.

▪ **Performance Characteristics:**

▪ The VVR provides linear output while

▪ Accumulators follow nonlinear gas laws (isothermal or adiabatic).

18

18

2.2.4- Standard (Industrial) versus Custom (Mobile) Reservoirs

❖ **Limitations in Mobile Applications**
❖ **Design:**

▪ Space limitation.

▪ Dynamic motion of the equipment.

▪ Pump intake must be covered by oil in all phases of machine operation.

▪ Avoid cavitation.

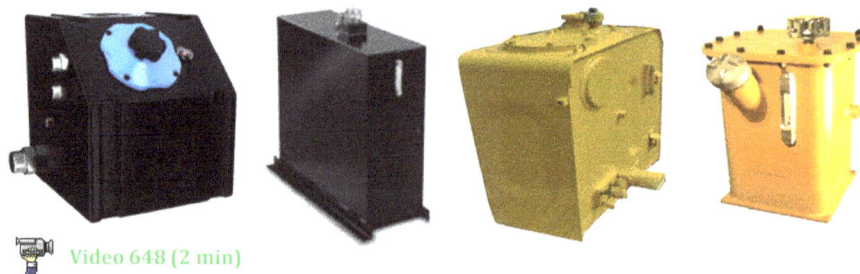

🎥 Video 648 (2 min)

Fig. 2.11 – Reservoirs for Mobile Applications

19

19

❖ **Shape:**
▪ Special dimensions → irregular shape.

Fig. 2.12 – Reservoirs for Mobile Applications (https://helgesen.com)

20

20

❖ **Material:**
▪ Reservoirs for mobile applications are made of different materials.

Steel Tanks Polyethylene Tanks Aluminum Tanks

Transfer Tanks Reefer Tanks

Fig. 2.13 – Reservoirs for Mobile Applications (www.americanmobilepower.com)

Upright Tanks Sidemount Tanks Saddlemount Tanks

21

21

2.2.5- Main versus Add-On Reservoirs

- Space limitations + large volume of oil is required →

- Add-On reservoir
o Is an extension to the main reservoir.
o Duration the oil stays in the reservoir ↑.
o Heat dissipation area ↑ + Dirt settlement ↑ + Air removal ↑.

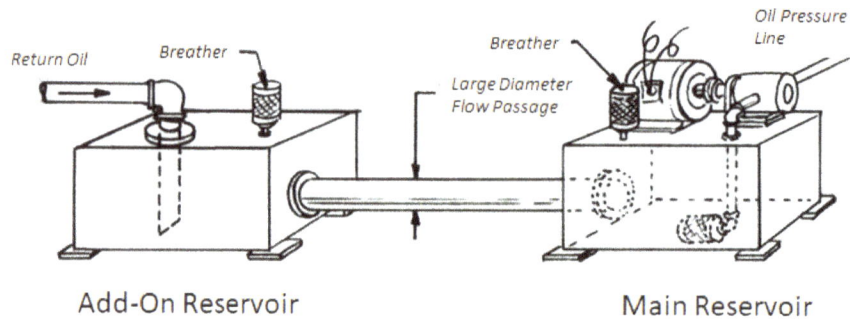

Fig. 2.14 – Main versus Add-On Reservoirs (Courtesy of Womack)

22

22

Best practices for setting up Add-On Tanks

- **Capacity:**
o Could be different from the main tank.
o Should consider the volume of return oil.

- **Suction:**
o Pump sucking oil from the main tank.

- **Return:**
o Return flow into the add-on tank.

- **Venting:**
o Each tank must be independently vented to atmosphere through a breather.

23

23

- **Connecting Line between Main and Add-On Tanks:**
  o Must connect both tanks at the lowest points.

- **Size of Connection Line:**
  o <u>Rule-of-thumb 1:</u>
  o Sized to keep flow velocity to 1 foot/s.

  o <u>Rule-of-thumb 2:</u>
  o Sized as (1 in^2 area)/(each 3 GPM of pump flow).
  o Example: if Q_p = 30 GPM→ connection line area = 10 squared inches.

24

24

2.2.6- Traditional (Steel) versus Modern (Hybrid) Reservoirs

Modern technology → "Hybrid Reservoirs or Tanks"

❖ **Product Description:**
- Ready-to-install complete module.
- All required tank functions are already integrated.

Fig. 2.15 – Hybrid Integrated Tank (Courtesy of Argo-Hyots)   25

25

❖ **Production Technology:**
- Material: High thermal strength Polyamide material.
- Production: Injection molding.

❖ **Connections with other Components:**
- Quick-Connect fittings → fault-free and tool-free hose connection.
- Filter housing is part of the tank → no sealing points → no risk of leakage.
- Easy filter element replacement.

❖ **Technical Data:**
- Tank Volume: up to 150 Liters.

- Temperature:
  - Range = (-30 to +100) $^{\circ}$C.
  - High thermal strength.

- Fluids:
  - Mineral oil and Environmentally Friendly hydraulic fluids.

26

26

❖ **Construction:**

1 Filter housing is integrated in the tank
2 Ventilating filter
3 Filling filter
4 Integrated oil level indicator
5 Quick-Connect fittings (see figure below)
6 Suction strainer
7 Baffle wall (in the shape of a channel)
8 Internal suction or return pipes
9 Sensor connections can be integrated
10 Oil drain plug

"Quick-Connect" system technology

Fig. 2.16 – Construction of Hybrid Integrated Tank (Courtesy of Argo-Hyots) 27

27

Video 522 (1 min)

❖ **Design Features:**

▪ Weight reduction → Meet weight limitations in mobile applications.

▪ Complex geometries → Meet space limitations in mobile applications.

▪ Special material → High mechanical & thermal strength.

▪ Special material → Excellent corrosion resistance.

▪ Production → No risk of leakage between the filter head and the tank.

▪ Construction → Cost savings compared to traditional tank solutions.

▪ Multiple ports → Flexible plumbing.

▪ Quick-Connect fittings → Tool-free assembly.

**Fig. 2.17 – Complex Shapes of Hybrid Integrated Tank
(Courtesy of Argo-Hyots)**

28

28

2.2.7- Reservoirs for Compact Power Units

▪ **Construction:**

o Unlike hybrid reservoirs (reservoirs + mounting block + components).

o The pump is immersed inside the reservoir.

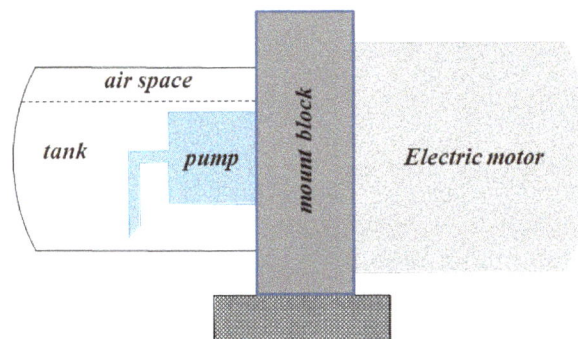

Fig. 2.18 – Concept of Compact Power Unit

29

29

- **The prime mover:**
  o It could be pneumatic, DC electric (12-24 Volt DC) or AC electric (one- or three-phase) type depending on the requirements.

- **The Pump:**
  o Standard extremal gear ($Qp = 1/10$ to 10 liter/min).
  o Some special versions can sustain pressures even higher than 1000 bar.
  o Some are controlled remotely to select pressure/flow.

- **Reservoir Material:**
  o Tanks are made of plastic or a light alloy with external cooling fins.

30

30

❖ **Applications:**
- Mobile applications + standalone hydraulic axes.
- Examples: civil automations for gates, doors, lifts, dentist's reclining chairs, pleasure boats, firefighting equipment, mobile machines for small forklifts, tailboards and dump bodies of self-propelled machines, machine tools and light presses.

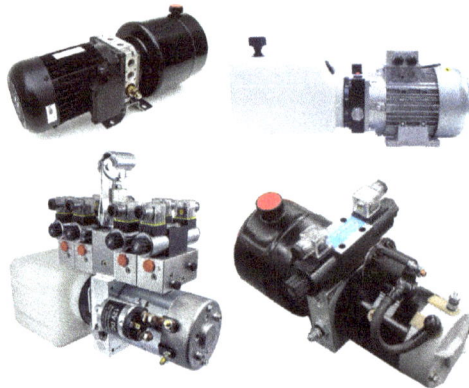

Fig. 2.19 – Examples of Compact Power Unit

31

31

Example (CytroPac Hydraulic Power Unit):

❖ **Construction:**

1. Oil tank with motor-pump group (optional cooling packages).
2. Central plate (integrated cooler).
3. Return flow filter.
4. Filter contamination sensors.
5. Filling level and temperature sensor.
6. Filling and breathing filters.
7. Cover with frequency converter below.
8. Electrical connections (see Fig. 2.21).
9. Visual oil level check and hydraulic fluid draining.
10. Clip (for removal of the hydraulic fluid hose for hydraulic fluid draining).

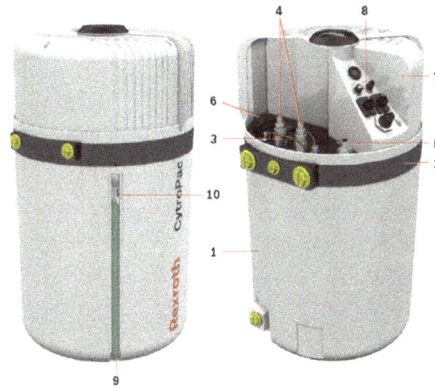

Fig. 2.20 – CytroPac Compact Power Unit (Courtesy of Bosch Rexroth)

❖ **Fields of Applications:**

▪ Machine tools and assembly lines under limited space conditions.
▪ Serve hydraulic power for continuous or intermittent demand.

32

❖ **Circuit Diagram:**

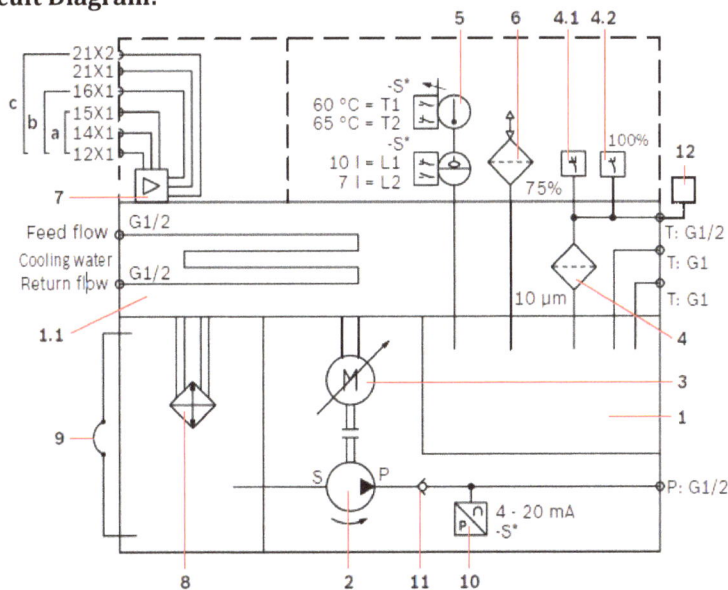

Fig. 2.21 – Circuit Diagram for CytroPac Compact Power Unit (Courtesy of Bosch Rexroth)

33

1	Oil tank			
1.1	Central plate			
	(integrated heat exchanger)			
2	Pump	7	Frequency converter
3	Motor	8	Cooling package (optional)
4	Return flow filter	9	Visual oil level check and
4.1	Filter contamination sensor 75%		hydraulic fluid draining
4.2	Filter contamination sensor 100%	10	Pressure load cell
5	Filling level and temperature sensor	11	Check valve
6	Filling and breathing filters	12	Filling coupling (optional)

34

34

❖ **Design Features:**

As reported by the unit's manufacturer:

- **Low Noise Level:** Plastic enclosure noise subrission.

- **Cost-Effective Operation:**
  o Frequency converter → power/speed regulation → operation costs ↓.

- **Various Configurations:**
  o Basic Configuration: sensors must be wired by the customer.
  o Advanced Configuration: sensors are wired by the manufacturer.
  o Premium Configuration: Condition is shared on Ethernet interfaces.

35

35

- **Prestart Control:**
  - Working pressure < max
  - → Unit is accelerated before actuators are connected

- **Sleep Function:**
  - Working pressure = max (e.g. during accumulator charging operation)
  - → Unit is automatically switched off. → energy efficiency ↑.

- **Warning Signals:**
  - Low oil level.
  - High temperature.
  - Filter clogging.
  - Frequency converter overload.

36

36

- **Characteristics:**
  - 1.5 - 4 kW (2 – 5.36 HP).
  - Various size pumps (4 – 14 ccm).
  - 4 ccm pump size → Pmax = 240 bar as long as Qp <10 l/min.
  - Power (Pmax = 240 and Qp =10 l/min) = 3.6 kW.
  - The unit keeps 0.4 kW power reserved for pressure spikes during start.
  - Q demand > 10 l/min → P ↓ accordingly to meet power requirements.

37

37

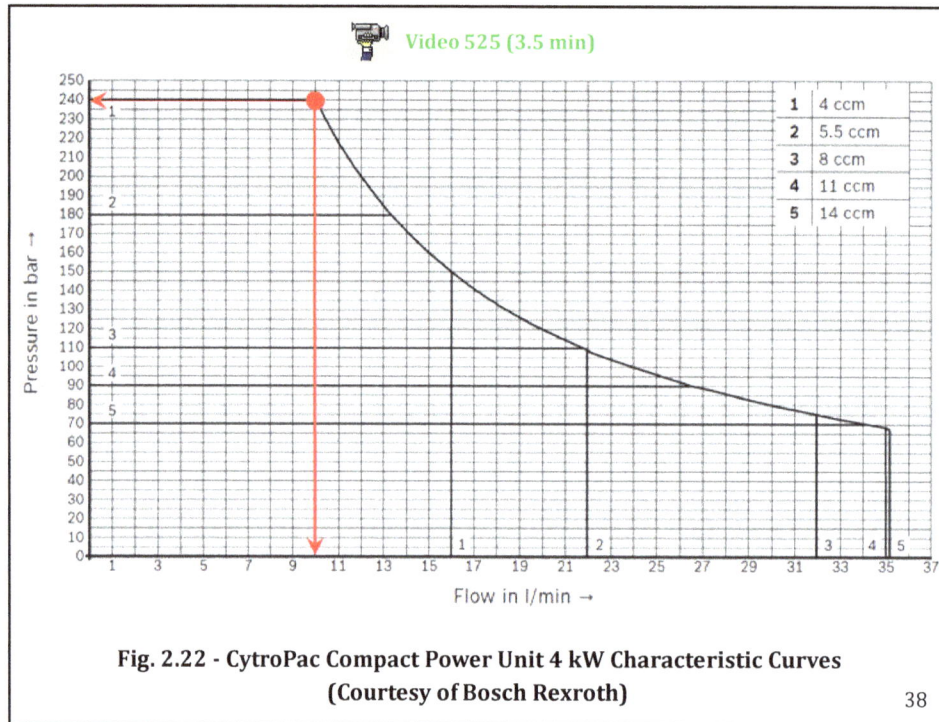

Video 525 (3.5 min)

1	4 ccm
2	5.5 ccm
3	8 ccm
4	11 ccm
5	14 ccm

**Fig. 2.22 - CytroPac Compact Power Unit 4 kW Characteristic Curves
(Courtesy of Bosch Rexroth)**

38

38

2.2.8- Foot-Mounted Reservoirs for Small Size Power Units

- The reservoir is placed on the ground and stands on two feet.
- Most common for small size industrial hydraulic power units.
- The upper cover is commonly used as a base for the equipment.
- Immersed pump (1) → better cooling & difficult maintenance.
- Alternatively, pump-motor unit can be assembled on top (2).

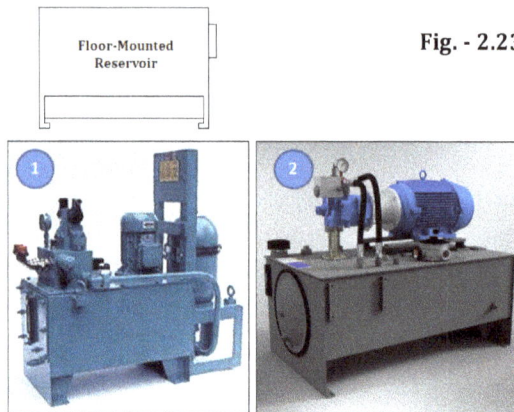

Floor-Mounted Reservoir

Fig. - 2.23 Foot-Mounted Hydraulic Reservoirs

39

39

2.2.9- L-Shaped Reservoirs for Medium Size Power Units

- Is recommended for medium size industrial hydraulic power units.
- Equipment are too heavy to be placed on top of the reservoir.

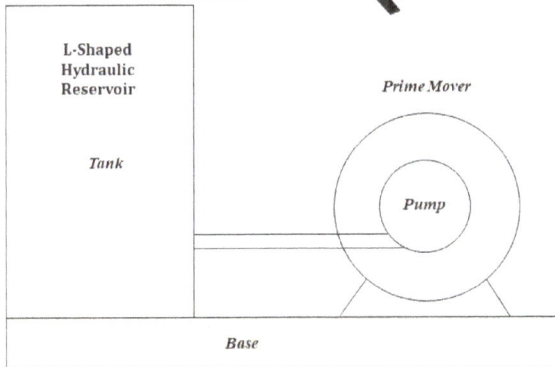

Fig. 2.24 - L-Shaped Mounted Hydraulic Reservoirs

40

2.2.10- Overhead Reservoirs for Large Size Power Units

- Overhead Reservoir → enables gravity flow → avoid cavitation.

- Positive P at the pump inlet (1 psi @2.5 ft for a typical petroleum oil).

- Recommended for pumps with large Q and/or high RPM.

- A typical application: large hydraulic presses and steel mills.

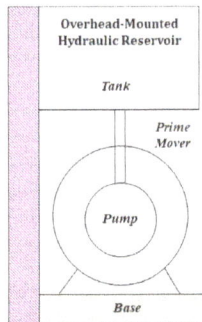

Fig. 2.25 - Overhead-Mounted Hydraulic Reservoirs

41

2.3- Construction of Hydraulic Reservoirs

- The reservoir is part of the Hydraulic Power Unit.
- Steel box + baffle plate + deaeration screen + cleaning holes + Accessories.
- Possible Accessories: accumulator + Heat exchanger + Off-line filtration unit.

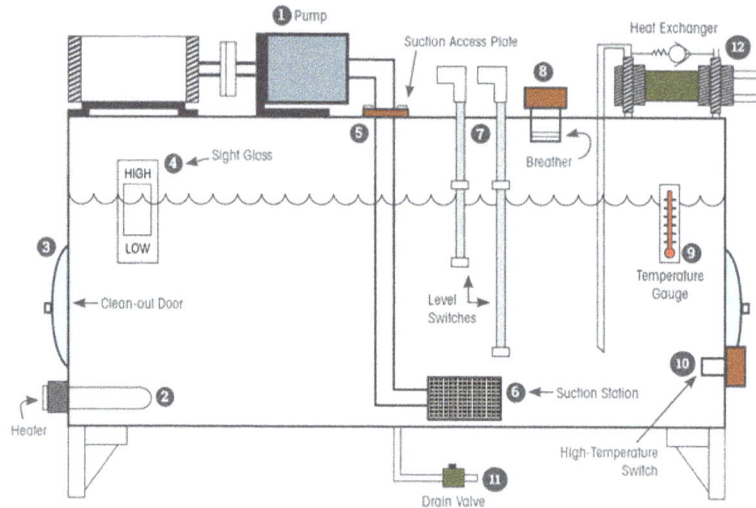

Fig. 2.26 – Construction of Hydraulic Reservoirs (www.efficientplantmag.com) 42

42

2.4- Design of Hydraulic Reservoirs for Industrial Applications

❖ **Improper design of a hydraulic reservoirs →**
- Oil aeration and foaming.
- Pump cavitation.
- Heating up the System.
- Building fluid contamination.

❖ **Design Standards:**
- SAE Aerospace Standard (AS5586).
- NFPA -T3.16.2-1969.
- ANSI -B93.18-1973 (R1987).
- ISO 4413:2010(E) 16 © ISO 2010.

❖ **Mobile** Applications → more complex + require modeling and simulation for vehicle dynamics during all phases of operation → out of scope of this textbook.

43

Design Considerations for Industrial Applications:
- **Body Construction:**
  1. Sizing.
  2. Heat Dissipation.
  3. Baffle Plates.
  4. Air Removal.
  5. Cleaning Holes.
  6. Drainage and Dirt Collection.
  7. Corrosion Protection.
  8. Structural Integrity.

- **Hydraulic Lines**
  9. Suction Lines.
  10. Return Lines.
  11. Drain Lines.

- **Accessories and Attachments**
  12. Filling Caps.
  13. Air Breathers.
  14. Oil Level and Temperature Indicators.

44

44

2.4.1- Sizing of Hydraulic Reservoirs

❖ **Maximum System Requirement (Vsmax):**

- Cylinders actuation + accumulators charging/discharging →
- Oil volume in the reservoir varies dynamically.

- **Vsmax =** Maximum oil volume required by the system (to extend all cylinders and charge all accumulators)

Video 388 (4.5 min)

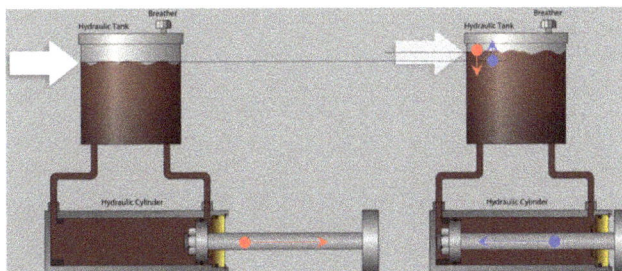

**Fig. 2.28 - Oil Level Change in the Reservoir during Machine Operation
(Courtesy of CD Industrial Group Inc.)** 45

45

❖ **Reservoir Filling Oil Volume:**

▪ **For Open Hydraulic Circuits:**

○ No restrictions on the space → large oil volume → settling time in the reservoir ↑ → better cooling, deaeration and contamination settlement.

○ Rule of thumb → **Vfill = (2-3) Qp.**   2.1A

○ **Note:** Vfill INCLUDES Vsmax

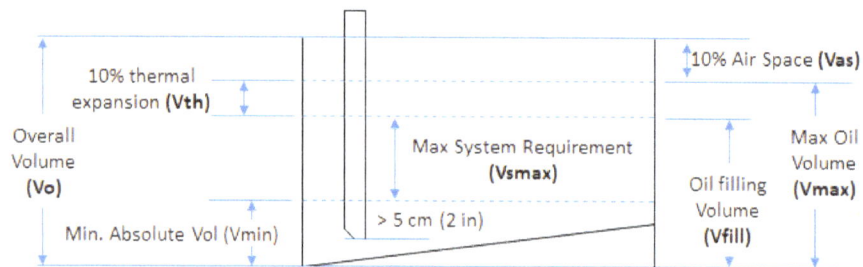

Fig. 2.27 - Rules for Sizing a Hydraulic Reservoir

46

▪ **For Closed Hydraulic Circuits (Hydrostatic Transmissions):**

○ Closed circuit →

○ Reservoir is sized based on the charge pump (not the main pump) →

○ Charge pump circulates case drains between the low-pressure side of the closed circuit and the reservoir →

○ Leakage from main pump and motor = (10-20)% of the main pump →

○ Rule of thumb → **Vfill = 0.2 x Qp.**   2.1B

47

❖ $Vth = 0.1 \times Vfill$ 2.2

❖ $Vmax = Vfill + Vth$ 2.3

❖ $Vas = 0.1 \times Vmax$ 2.4

❖ Reservoir Overall Volume (Vo) = $Vmax + Vas = Vfill + Vth + Vas$ 2.5

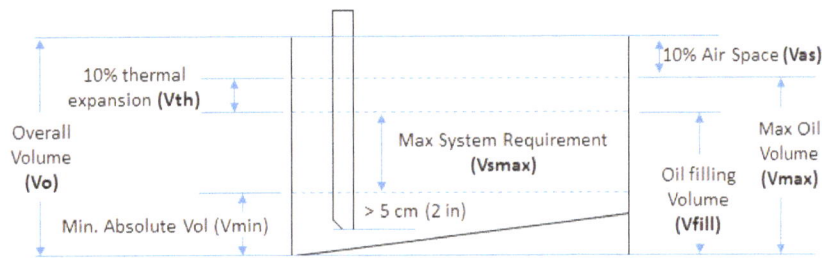

Fig. 2.27

48

48

❖ **Absolute Minimum Oil Volume (Vmin):**

Absolute Minimum Oil Volume (Vmin) = Vfill - Vsmax   2.6

▪ **Checks on Vmin:**

o Allow sufficient fluid supply & Avoid pump cavitation.

o Cover the suction line by at least 5-10 cm (2-4 inches) when all cylinders are in their extended positions and accumulators are fully fluid charged .

o Consider the reservoir inclination in mobile applications.

49

49

2.4.2- Heat Dissipation from Hydraulic Reservoirs

- A reservoir → passive system cooling.

- If passive cooling is not sufficient → active cooling using heat exchanger.

- Design considerations to maximize the heat dissipation from a reservoir to the surrounding area.

A. Shape and Dimension of the Reservoir.
B. Reservoir Ground Clearance.
C. Material of the Reservoir.
D. Cooling Fins.

50

A. Shape and Dimension of the Reservoir:

- **Reservoir Overall Volume (Vo)**
  $$= H \times W \times L = 0.85W \times 0.85L \times L = 0.85^2 L \times 0.85L \times L = (0.85 \times L)^3 \qquad 2.7$$
- If this calculation results in nonstandard reservoir length →
- Consider the nearest standard length.
- **Reservoir Width (W) = 0.85 L** $\qquad$ 2.8
- **Reservoir Footprint Area (AR) = L x W** $\qquad$ 2.9

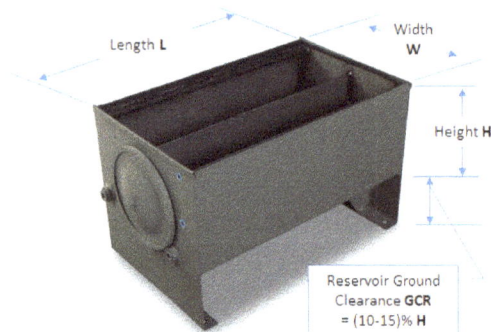

Fig. 2.29 - Recommended Shape and Dimensions of a Reservoir for better Heat Dissipation

51

- **Minimum Oil Height (Hmin) = Vmin/AR**   2.10
- **Maximum Oil Height (Hmax) = Vmax/AR**   2.11
- **Reservoir Height (H) = Vo/AR**   2.12

B. Ground Clearance of the Reservoir (GCR):
- Ground clearance → allow air flow → better heat dissipation.
- **Reservoir Ground Clearance (GCR) = 0.15 x H**   2.13

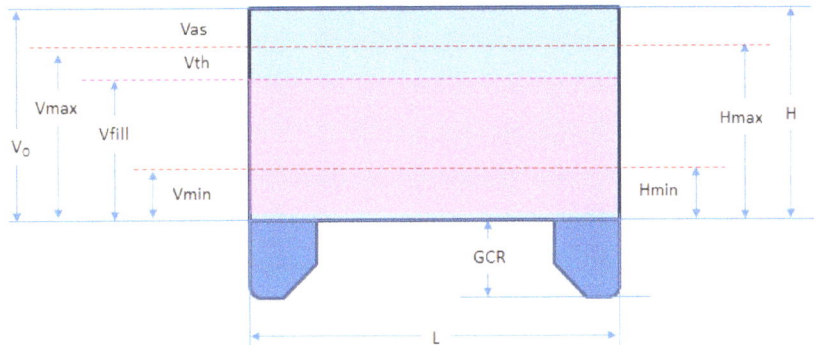

Fig. 2.30 - Fluid Hight in the Reservoir during Machine Operation   52

52

C. Material of the Reservoir:
- Materials of higher heat transfer coefficient (Steel or Aluminum).

D. Cooling Fins: Video 246 (0.5 min)

Fig. 2.31 - Cooling Fins Welded to the Side Walls to Improve Cooling Capacity
(Courtesy of Womack)   53

53

E. Calculation of the Cooling Capacity (CC):

- Heat is dissipated from reservoir walls and bottom.
- As a safety factor →
- Consider side walls only and **Hmin** (or average H depends on operation)

- **Contact (WET) Area (AC) = 2 x Hmin x (L + W)**　　　　2.14
- **Cooling Capacity (kW) = AC (m^2) x ΔT ($^{\circ}$C) x K**　　2.15A
- **Cooling Capacity (HP) = AC (ft^2) x ΔT ($^{\circ}$F) x K**　　2.15B

Where:
- **1 HP = 2545 BTU/HR**
- **1 kW = 3413 BTU/HR**
- **1 kW = 1.341 HP (i.e. 1 HP = 0.746 kW)**
- **ΔT = Differential temperature (Fluid Temp. – Air Temp.).**
- **K = Heat Transfer Factor.**
- **For Eq. 2.15A:**
- Stainless Steel (K = 0.0058), Carbon Steel (K = 0.0144), and Aluminum (K = 0.0693).
- **For Eq. 2.15B:**
- Stainless Steel (K = 0.0004), Carbon Steel (K = 0.001), and for Aluminum (K = 0.0048).

54

54

2.4.3- Baffle Plates
- Baffle Plate → divides reservoirs → separates return line from suction line → Oil settling time in the reservoir ↑→ Better cooling, deaeration, and contamination settlement.
- **Baffle Plate Height (BPH) is slightly > Hmax.**
- **BPH (cm) = Hmax (cm) + (1-5) cm**　　　　2.16

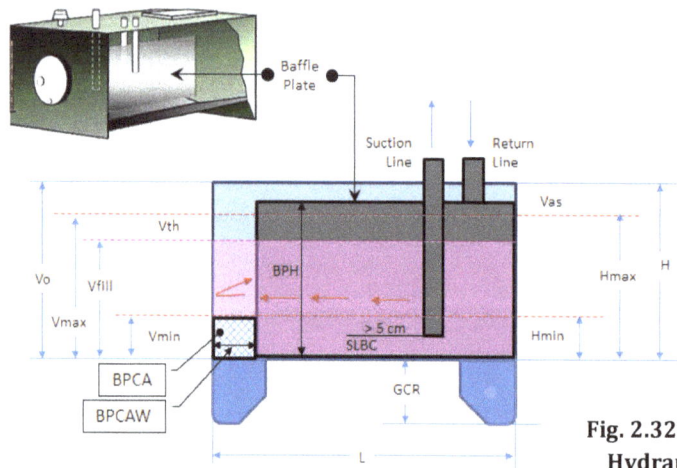

Fig. 2.32 - Baffle Plate in Hydraulic Reservoir　55

55

- **Baffle Plate Connecting Area (BPCA)**
- Rule of thumb → (1 cm^2 for every 2 lit/min) OR (1 in^2 for every 3 GPM).
- **BPCA (cm^2) = Qp (lit/min)/2   2.17**
- **Width of Baffle Plate Connecting Area (BPCAW) (cm)**
  = BPCA (cm)/Hmin (cm)   **2.18**

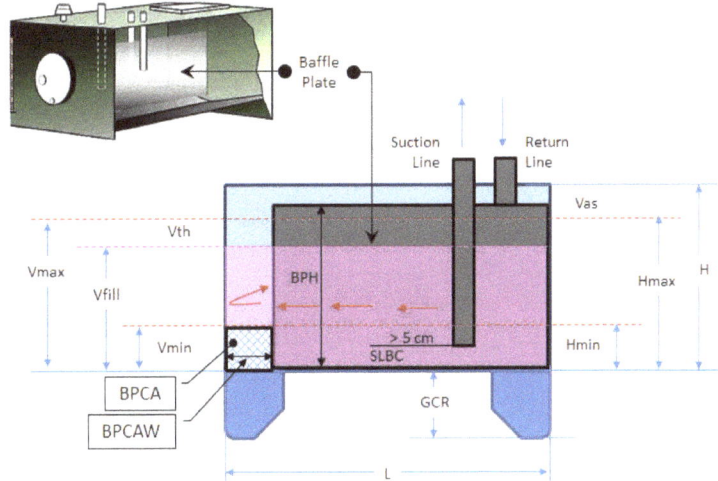

56

2.4.4- Air Removal

- 100-micron mesh size screen.
- The screen is placed inside the reservoir with 30-45 degrees.
- Return oil arrives at the upper side of the screen →
- Air bubble gathered → bigger bubbles are formed →
- Bigger bubbles float easily upwards and dissipate at the surface.

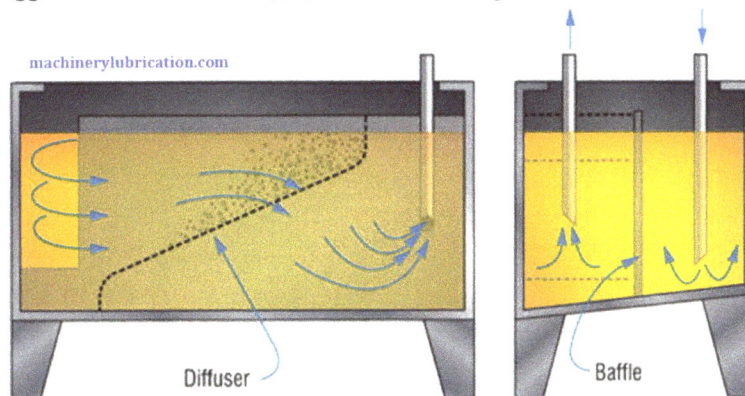

Fig. 2.33 - Air Removal Screen in Hydraulic Reservoir   57

2.4.5- Cleaning Holes

- Cleaning Hole (Manhole).
- Large enough → maintenance person get inside.
- Gasket → prevent external leakage.

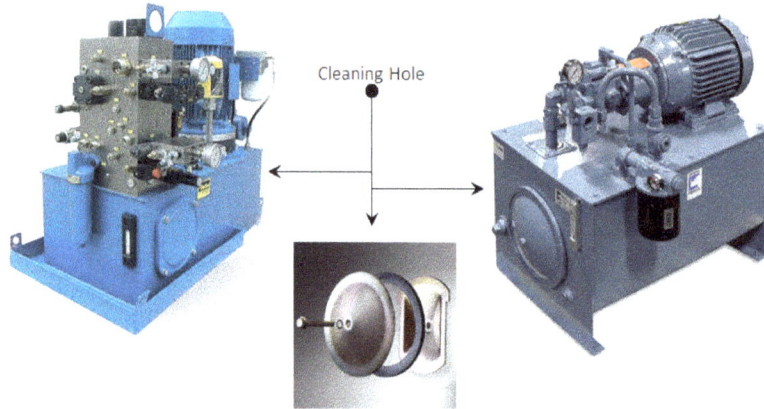

Cleaning Hole

Fig. 2.34 - Cleaning Hole in Hydraulic Reservoirs

58

58

2.4.6- Drainage and Dirt Collection

Reservoirs should be designed to allow:
- Complete oil draining in place.
- Easy removal (water and contamination).
- Easily access (strainers, diffusers & other internal components).

Considerations: Bottom of the reservoir, Drain Plugs , and Drain Valves.

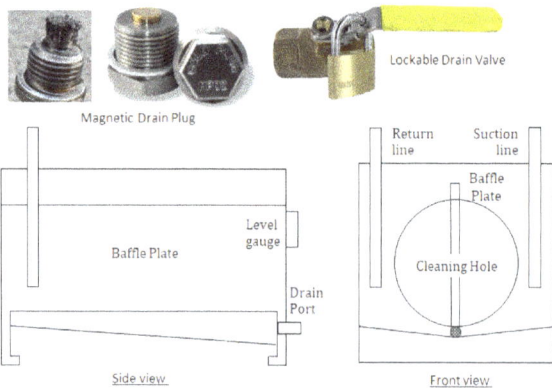

Magnetic Drain Plug

Lockable Drain Valve

Return line   Suction line

Baffle Plate

Cleaning Hole

Level gauge

Baffle Plate

Drain Port

Side view

Front view

Fig. 2.35 - Reservoir Design for Better Drainage and Dirt Collection

59

59

2.4.7- Corrosion Protection

- Steel Walls → inside rusting.

- Water-based hydraulic fluids → inside rusting.

- Water content in mineral fluids → inside rusting.

- Environmental conditions → outside rusting.

- → Reservoir inside and outside painting is required.

- Inside painting → compatible with hydraulic fluid.

- Phosphate Ester is not compatible with ordinary paint.

- If Phosphate Ester is used

- → previous incompatible paints must be stripped and replaced.

- Electrostatic Powder Spraying is recommended because it creates a hard-finished skin that is tougher than conventional paint.

60

60

2.4.8- Structural Integrity of Hydraulic Reservoirs

- **Pressure Change:**
  o Fluid withdrawal or return →
  o Reservoirs are exposed to positive and negative pressures →
  o Reservoir structure must have enough strength.

- **Weight of the Fluid/Components:** Must be considered when determining the strength of the reservoir body.

- **Lifting:**
  o Lifting points (distribution of components & CG).
  o Forklift usage carrying option may be considered.

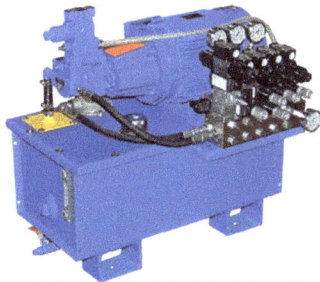

**Fig. 2.36 – Forklift Carrying Option
(hyvair.com)**

61

61

2.4.9- Suction Lines

❖ **Suction Line Diameter DS:**
- Most critical line in a hydraulic circuit.
- Undersized suction line → pump cavitation →

- Guidelines for sizing based on:
- 1-Pump Suction Port: for a given pump size and driving speed.

- 2-Flow Speed in the Line:
- Rule of thumb → flow speed**(v)** = (0.6-1.2) m/s = (2-4) ft/s.
- → minimize the pressure losses and avoid pump cavitation.
- Pump flow **(Q)** and suction line crossection area **(A).**

$$Q\left[\frac{\text{lit}}{\text{min}}\right] = \frac{v\,[\text{cm/s}] \times A\left[\text{cm}^2\right] \times 60}{1000} \qquad 2.19A$$

$$Q[\text{gpm}] = \frac{v\,[\text{fps}] \times A\left[\text{in}^2\right]}{0.321} \qquad 2.19B$$

- 3-Reynold's Number:
- Reynold's Number **Re** < 2000 → Laminar Flow → avoid cavitation.
- Pump flow **(Q)**, and fluid viscosity **(v).**

$$D[\text{mm}] = \frac{21231\,Q\left[\frac{l}{\text{min}}\right]}{v\,[\text{Cst}] \times R_e} \qquad 2.20A$$

$$D[\text{mm}] = \frac{3164\,Q[\text{gpm}]}{v\,[\text{Cst}] \times R_e} \qquad 2.20B$$

o 4-Validation using HCSC Software:

o https://www.compudraulic.com/software

o Example: For a pump that discharges 100 lit/min (25 gpm).

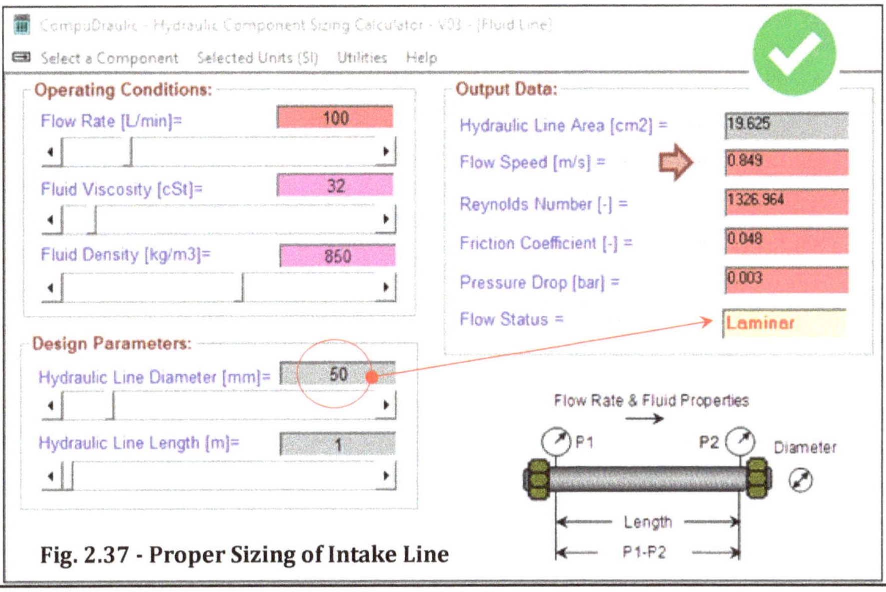

Fig. 2.37 - Proper Sizing of Intake Line

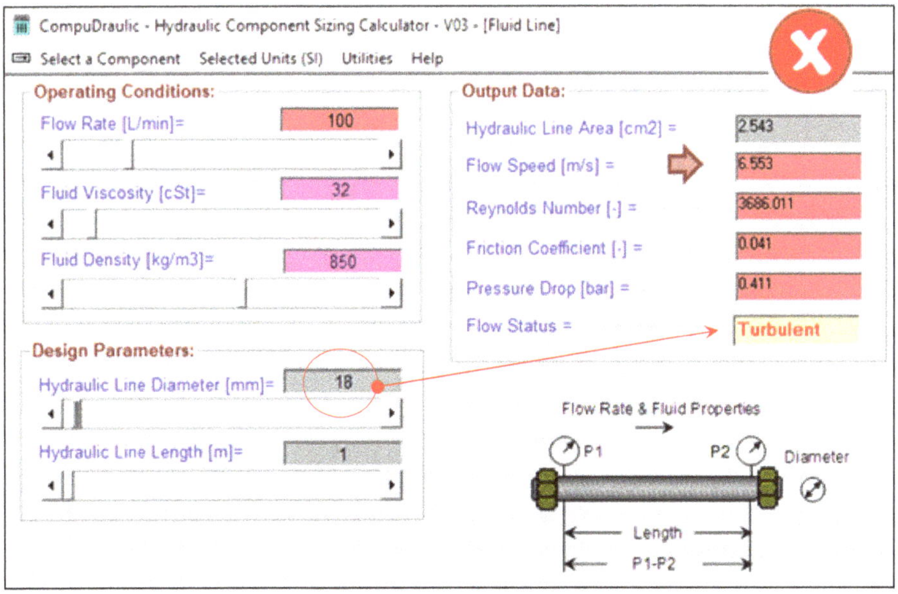

❖ **Suction Line Placement:**
- Distance between suction line and return line ↑.
- Distance between suction line and point of dirt collection↑.

Video 643 (1.5 min)

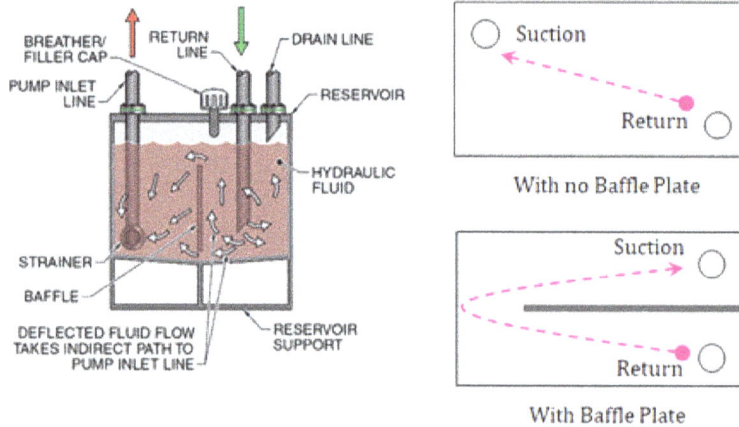

With no Baffle Plate

With Baffle Plate

Fig. 2.38 - Suction Line (Courtesy of American Technical Publisher)

66

66

❖ **Suction Line Bottom Clearance (SLBC):**
- Bottom is edged on 45° → enlarge the opening → reduce pressure losses.
- Entrance shall be directed to the wall → make it difficult to suck dirt.
- Suction Line Bottom Clearances Rule of thumb →

Suction Line Bottom Clearance (SLBC) = (2-3) x (DS)   2.21

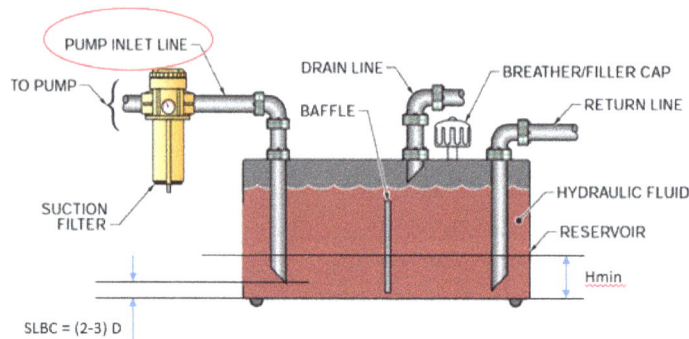

Fig. 2.38

67

67

❖ **Suction Line Covered Bottom (SLCB):**

▪ Rule of thumb → SLCB > 5-10 cm (2-4 in).

▪ Otherwise, minimum oil volume should be adjusted accordingly.

Suction Line Covered Bottom (SLCB) = Hmin - SLBC　　　　2.22

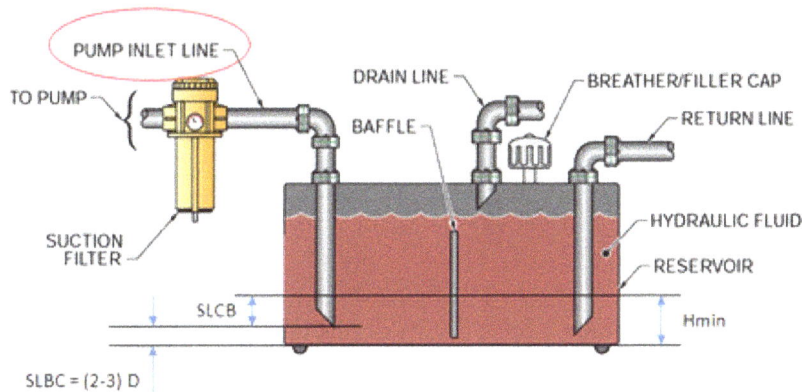

Fig. 2.38

68

68

❖ **Suction Line Length Ls:**

▪ Very long → more pressure losses → pump cavitation.

▪ Very short → turbulence in fluid transferred into the pump intake port.

▪ **Rules of thumbs:**
▪ **First:** Suction Line Length
▪ **Ls = 10 DS**　　　　　　　　　　　　　　　　　　　2.23

▪ **Second:**
▪ **Ls >= 25 cm (10 inches)** starting from last turbulent point (such as a strainer or a bend in the line) → flow get back as laminar before the pump port.

▪ **Third: Ls** < maximum length that develops maximum permissible pressure losses in the line.

Note:
▪ Rule 2 Overrides Rule 1.
▪ Rule 3 Overrides Rules 2.

69

69

- **Example 1:**
  - Assuming applying 1st rule → L_S = 10 cm and
  - Assuming applying 2nd rule → L_S >= 12 cm,
  - Then so far, L_S >= 12 because second rule overrides the first one.
  - Otherwise, flow might be turbulent at the intake port.
  - Assuming applying 3rd rule → L_S <= 15 cm.
  - Then (12 cm >= L_S <15 cm)
  - Otherwise, pressure drop in the line increases above the permissible.

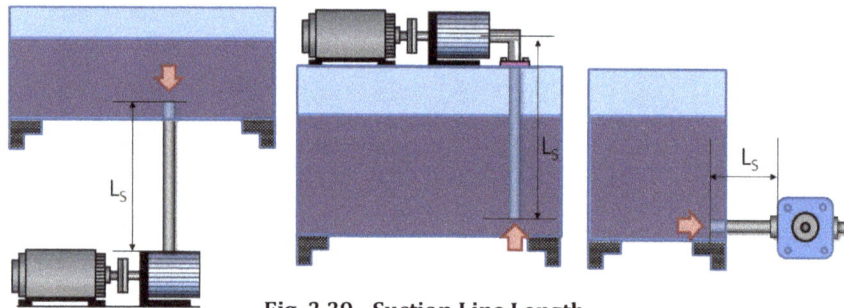

Fig. 2.39 - Suction Line Length

70

70

- **Example 2:**
  - Assuming applying 1st rule → L_S = 15 cm and
  - Assuming applying 2nd rule → L_S = 13 cm,
  - Then so far L_S >= 13
  - This complies with both rule 1 & 2.
  - Assuming applying third rule → LS <= 10 cm.
  - Then then LS <=10 cm
  - Otherwise, pressure drop in the line increases above the permissible.

71

❖ **Suction Head and Suction Pressure:**

- ± suction head (**Hs**) depends o the pump altitude from the reservoir.
- Positive suction head → positive intake P → avoid cavitation.
- Negative suction head → negative intake P → possible cavitation.
- Pump manufacturer → maximum permissible **NSP**.
- If no information → (NSP) >= -0.2 bar gauge (-3 psig).

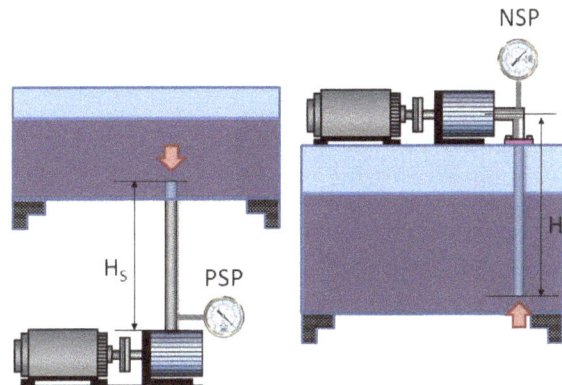

Fig. 2.40 - Suction Head and Suction Pressure 72

72

$$Hs\ (m) = \frac{SP\ (Pa)}{SG \times \gamma_W (N/m^3)} = \frac{10^5 \times SP\ (bar)}{SG \times 9.81 \times \gamma_W (kg/m^3)}$$

$$\gamma_W \left(\frac{kg}{m^3}\right) = 1000 \rightarrow Hs\ (m) = \frac{100 \times SP\ (bar)}{SG \times 9.81} \qquad 2.24A$$

$$HS\ (ft) = \frac{SP\ (psi)}{SG \times \gamma_W (lb/ft)} = \frac{SP\ (psi)}{SG \times 62.4 \times \gamma_W (lb/lb^3)} = \qquad 2.24B$$

Where (Refer to Fig. 2.40):

- **Hs** = Suction Head

- **SP = PSP** = Positive Suction Gauge Pressure (corresponds to positive suction head when the pump is placed below the surface of the oil).

- **SP = NSP** = Negative Suction Gauge Pressure (corresponds to negative suction head when the pump is placed above the surface of the oil).

- γ_W = Specific Weight of water & **SG** = Specific Gravity of the hydraulic fluid. 73

73

❖ **Suction Line Type:**

- Hard tubing is highly recommended over flexible hoses.
- Hoses take misalignment between the pump and reservoir.
- Make sure a suction hose (NOT a pressure hose) is used.
- Suction hoses are specially constructed (spiral wire reinforcing inner layers due to negative pressure).

Fig. 2.41 - Example of Suction Hose Collapse

74

❖ **Suction Line Routing:**

- Straight intake line with minimum #bends
- → minimize the line losses.

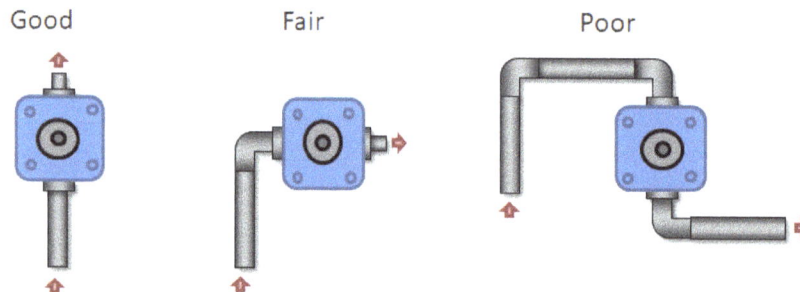

Good   Fair   Poor

**Fig. 2.42 - Example of Proper Intake line Routing
(Courtesy of Womack)**

75

❖ **Suction Strainer:**
- Strainers are not recommended for large flow pumps.
- It subject to pump manufacturer approval.
- **Micron Size:** Largest particle micron size can pass through the screen.
- **Mesh Size:** Number of holes in a 1 squared inch.
- **Micron Size vs. Mesh Size:** Mesh Size ↑→ Micron Size↓.
- **Mesh Size:** If no information were found → 250-500 is just fine.
- **Surface Area:**
- Consult the pump manufacturer.
- Rules of Thumbs:

Suction Strainer Surface Area (cm^2) = 3 x Qp (lit/min)   **2.25A**

Suction Strainer Surface Area (in^2) = 2 x Qp (gpm)   **2.25B**

76

76

- **Connection with Suction Line:**
- **Bypass Valve:**   Video 641 (0.5 min)
- Suction strainers are hidden
- → bypass valve and/or Δp indicator are recommended.

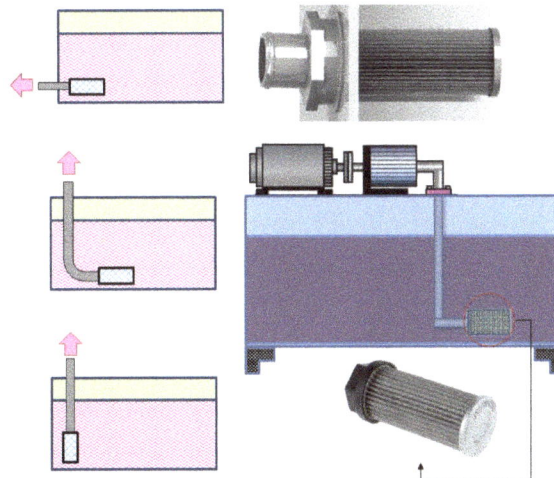

Fig. 2.43 – Suction Strainers

77

77

❖ **Shutoff Valve:**
- If a pump installed below the fluid surface → a shutoff valve is required.
- MUST BE lockable → avoid pump failure if it is accidently closed.

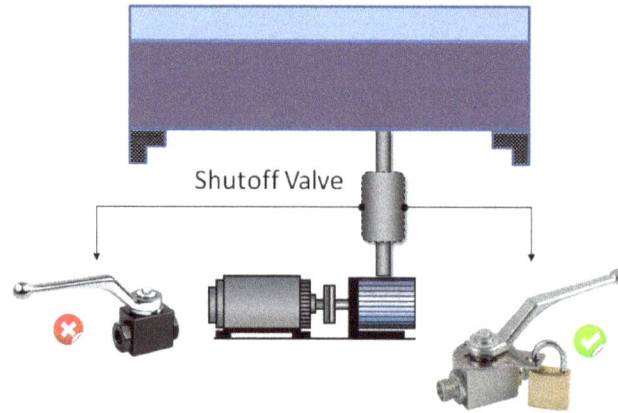

Fig. 2.44 – Shutoff Valve on Suction Line

78

78

❖ **Suction Pressure Measurement:**
- For pump suction condition monitoring → vacuum gauge.
- For pump protection → vacuum switch.

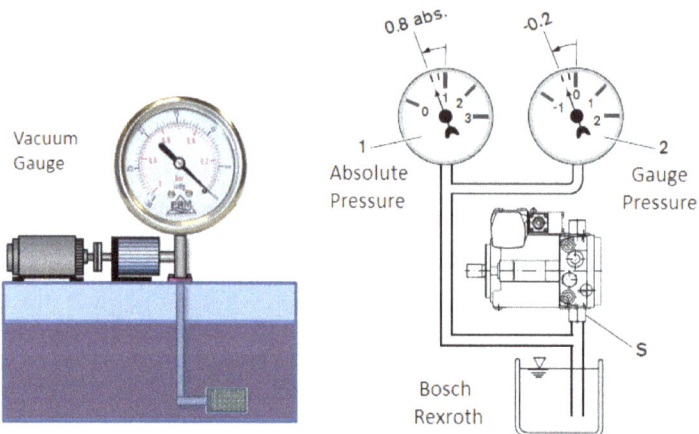

Fig. 2.45 – Pump Suction Pressure

79

❖ **Suction Line Pressure Boosting:**

- Open Circuits → Recommended for large size and high-speed pumps.
- Closed Circuits → Mandatory.

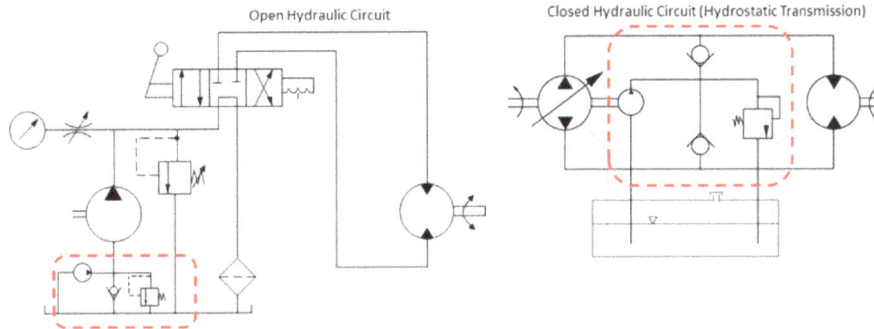

**Fig. 2.46 - Intake Line Pressure Boosting
in Closed and Open Hydraulic Circuits**

80

80

2.4.10- Return Lines

❖ **Return Line Placement:**

- In accordance with the placement of the suction line and the baffle plate.
- Edged end is directed against the reservoir wall to maximize oil cooling.

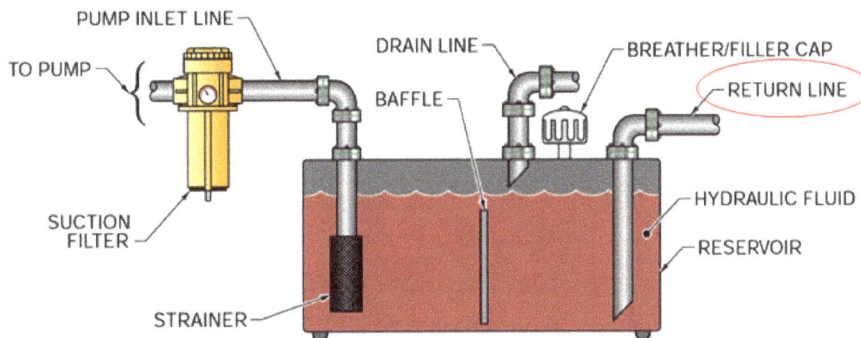

Fig. 2.47 –Return Line (Courtesy of American Technical Publisher)

81

81

❖ **Return Line Diameter DL:**

- Thinking that a return line is used just to bring oil back to the reservoir

- → NOT WISE.

- Undersized return line → back pressure + turbulence to the reservoir.

- Equations 2.19 and 2.20 are also applicable to determine **DL**:

o Rule of thumb → flow speed **(v)** = (1.2-2.1) m/s = (4-7) ft/s.

o A common mistake is based on a wrong assumption → return flow = **Qp**.

o Differential cylinders → Return flow > **Qp**.

o Accumulator discharging → Return flow > **Qp**.

o Flow distribution analysis is required before sizing a return line.

❖ **Return Line Bottom Clearance (RLBC) and Covered Bottom (RLCB):**

- Same rule of thumb that applied to suction line is applied for return line.

- Equations 2.21 and 2.22 are used to determine these parameters.

82

82

❖ **Diffusers:**
- A return line is edged on 45 degrees → acts like a diffuser →
- help fluid deaeration.
- Installing a diffuser → fluid aeration, foaming and noise are reduced.
- → Pump life is extended, and cavitation is avoided.

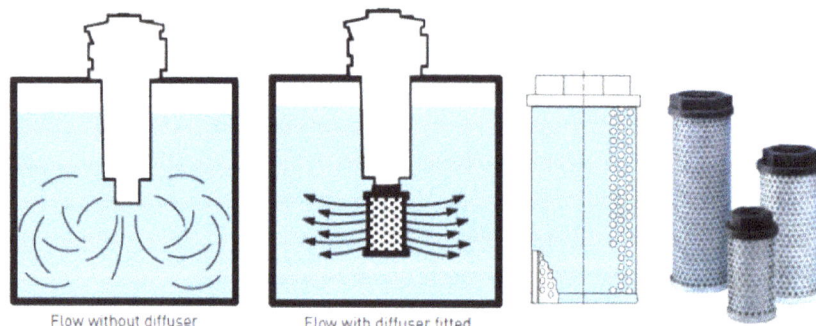

Flow without diffuser   Flow with diffuser fitted

Fig. 2.48- Diffusers on a return line (Courtesy of Parker)

83

83

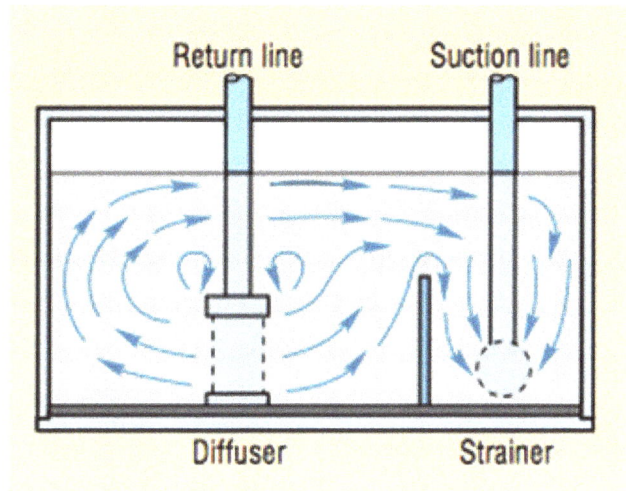

Fig. 2.49- Flow Stream from the Return Line to Suction Line (Courtesy of Parker)

84

84

2.4.11- Drain Lines

- Some of hydraulic valves (?) and pumps (?) have drain lines.
- Drain lines are separated from main return line
- → avoid main return back pressure.
- Drain flow can discharge either on top or underneath the oil level.

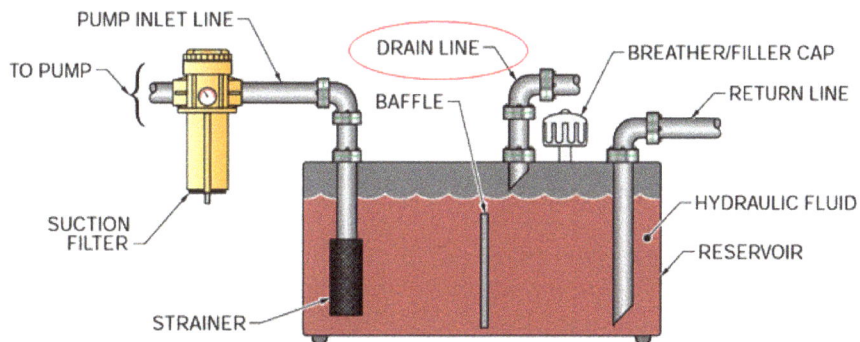

Fig. 2.50 –Drain Line (Courtesy of American Technical Publisher) 85

85

2.4.12- Filling Caps

- Sealed cover → less contamination ingression.
- Filling screen → catch relatively large contaminants during filling.
- Caps are chained to the reservoir → keep them captive.
- Lockable caps → more protection.

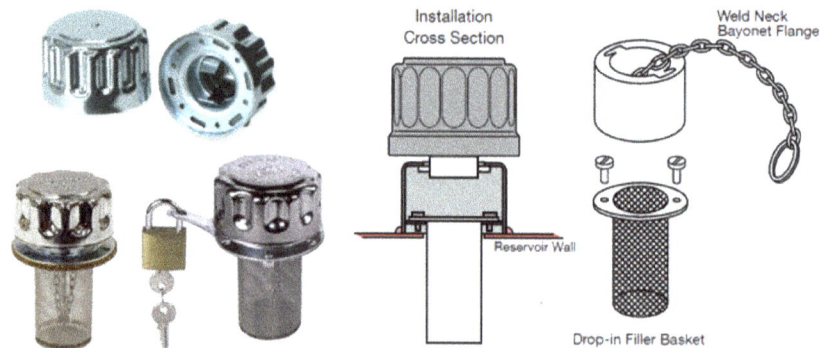

Fig. 2.51 –Filling Caps

86

2.4.13- Air Breathers

- Single-acting cylinder + differential cylinders + accumulators →
- Oil level in the reservoir dynamically changing →
- Reservoir breathes during the machine operation like a human.
- Air contains contaminants, dust, and moisture →
- Air breathers are required → prevent ingression of these things.

❖ **Standard Filler Breathers:**

- They are available in various forms and sizes.
- Metallic or non-metallic, flange-mount or thread-mount.
- Very similar in shape to the filler caps.
- Other styles can look like a spin-on oil filter.
- 10 microns size of a conventional or telescopic strainer.

87

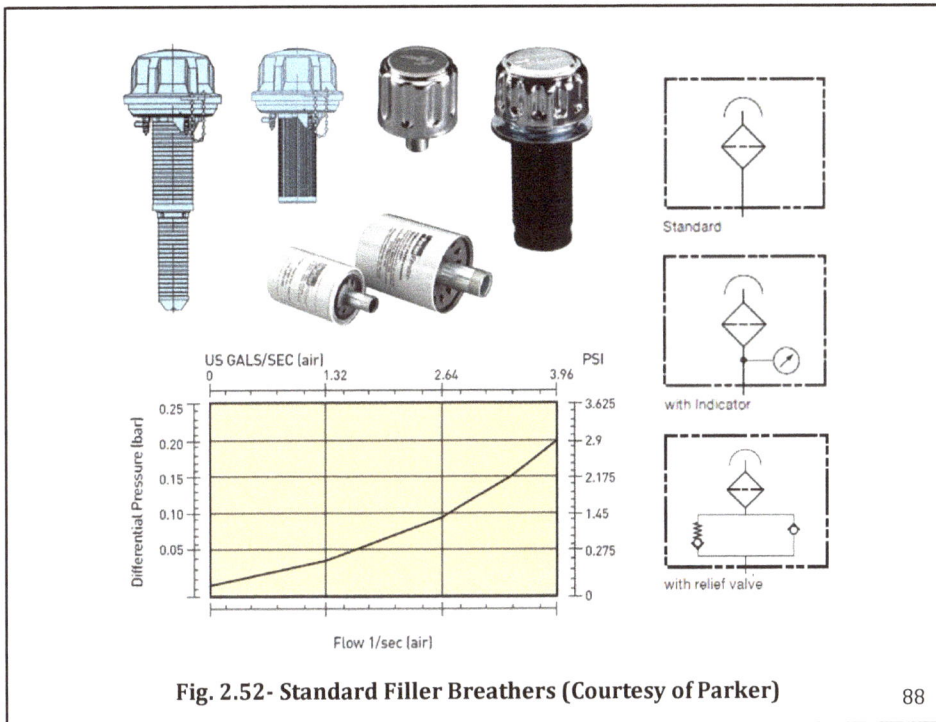

Fig. 2.52- Standard Filler Breathers (Courtesy of Parker)   88

88

❖ **Desiccant Breathers:**
- Highly humid environment → desiccant breathers.
- Transparent body filled with silica-gel → absorb 40% of its weight.
- Color changes (blue to pink → saturated → replacement.
- Optional check valve on inlet → Prevent saturation during shutdown.
- Optional check valve on outlet → exhaust air from the tank bypass the gel.

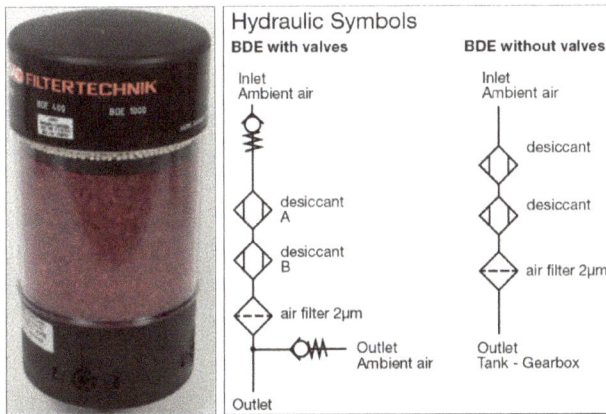

Fig. 2.53 - Breather Dryer (Courtesy of HYDAC)

89

89

❖ **Construction and Operation of Desiccant Breather:**
▪ Working T range -30 $^{\circ}$C to 100 $^{\circ}$C (-22 $^{\circ}$F to 212 $^{\circ}$F).

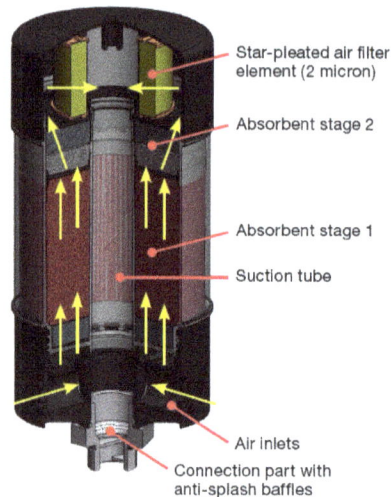

- Star-pleated air filter element (2 micron)
- Absorbent stage 2
- Absorbent stage 1
- Suction tube
- Air inlets
- Connection part with anti-splash baffles

Fig. 2.54 – Construction and Operation of the Desiccant Breather Dryer (Courtesy of HYDAC)

90

90

❖ **Breathers Dryers:**
▪ Water absorption cartridges.
▪ Collect and expel moisture out of reservoirs.
▪ Unlike Desiccant Breather, cartridge color won't change when saturated.

Video 406 (0.5 min)

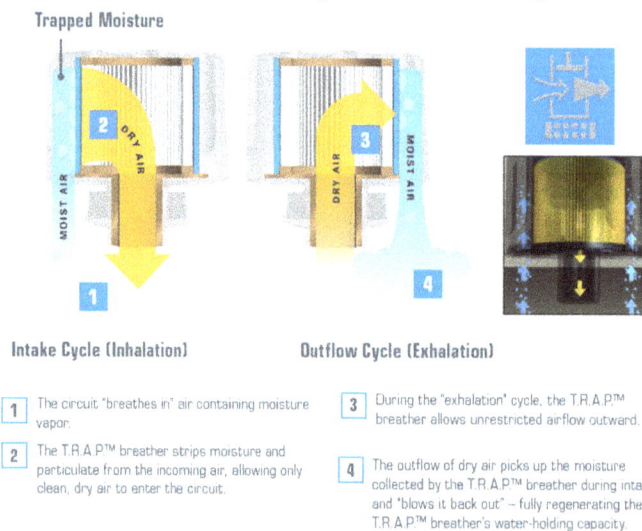

Trapped Moisture

MOIST AIR — 2 — DRY AIR — 1

DRY AIR — 3 — MOIST AIR — 4

Intake Cycle (Inhalation)

1	The circuit "breathes in" air containing moisture vapor.
2	The T.R.A.P™ breather strips moisture and particulate from the incoming air, allowing only clean, dry air to enter the circuit.

Outflow Cycle (Exhalation)

3	During the "exhalation" cycle, the T.R.A.P™ breather allows unrestricted airflow outward.
4	The outflow of dry air picks up the moisture collected by the T.R.A.P™ breather during intake, and "blows it back out" – fully regenerating the T.R.A.P™ breather's water-holding capacity.

Fig. 2.55 – Construction and Operation of Breather Dryers (Courtesy of Donaldson)

91

91

❖ **Sizing of Breather Dryer:**

- Larger size breather → better water retention → larger pressure drops.

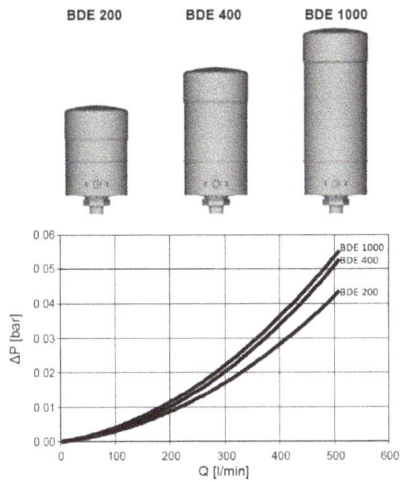

Size	Maximum water retention capacity
200	0.25 l
400	0.50 l
1000	0.75 l

Fig. 2.56 – Sizing of the Breather Dryer (Courtesy of HYDAC)

92

92

❖ **Air Breathers for Closed Tanks:**

- Highly contaminated environment → Closed (pressurized) reservoir →
- Conventional air breather can't be used →
- Air returns with the oil, accumulated on top of the oil surface.
- An air check valve is used to protect the tank against over-pressure.
- Cracking pressure of (1 – 3) PSI is usual.

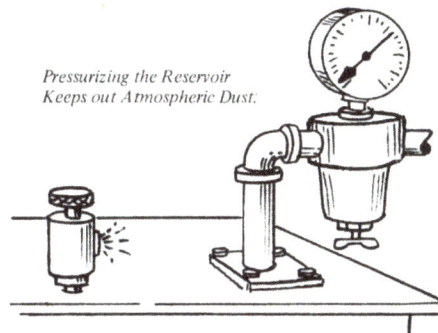

Pressurizing the Reservoir Keeps out Atmospheric Dust.

Fig. 2.57 – Breathers for Pressurized Reservoirs (Courtesy of Womack)

93

2.4.14- Oil Level and Temperature Indicators

❖ Basic Requirements:

▪ High/Low permanent marks.

▪ Installed so that they are clearly visible during filling.

❖ Optional Requirement:

▪ Switches & Sensors → monitoring, alarming, protection.

94

94

❖ **Sight Glasses:**
1. Standard rounded pod.
2. Oil color and clarity.
3. Oil aeration and foaming.
4. Varnish formation.
5. Observe corrosion.
6. Observe wear debris.
7. Sampling hydraulic fluid.

Video 642 (2 min)

**Fig. 2.58 – Multifunction
Sight Glass
(www.luneta.com)**

95

95

❖ **Oil Level and Temperature Indicators**:
- It is installed and sealed in drilled holes.
- Some contain a thermometer.
- Large tanks → two small level indicators.
- Mobile equipment → use a dipstick to check fluid level.

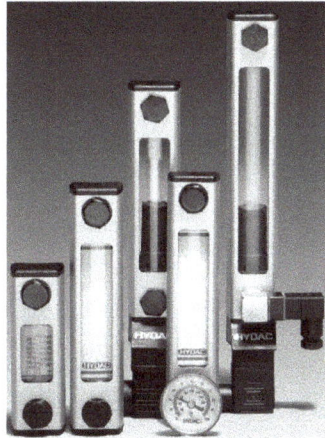

Fig. 2.59 – Oil Level and Temperature Gauges and Switches
(Courtesy of Hydac)

96

96

❖ **Oil Level and Temperature Switches**:
- Temperature switches → machine protection.
- Optimum range of working T = 38°C to 54°C (100°F to 130°F).
- Critical T for a typical Petroleum-Based hydraulic fluid is 70°C (158°F).
- Every 10°C (18°F) incremental increas above the critical T→
- Oxidation rate of the hydraulic fluid is doubled.
- Example: running a system at a consistent 80°C (176°F) →
- Fluid life is reduced by 75%.
- Set the high-temperature switch at 70°C (158°F) to shut off the pump and prevent oil breakdown.

97

97

2.5- Hydraulic Reservoir Design Case Study

Statement of the Problem: refer to the textbook.

Summary of the given data:

- Pump Flow **Qp** = 100 lit/min.

- Pump suction port = 1.5 in (3.81 mm).

- The pump withdraws the oil through a strainer.

- Fluid Viscosity **v** = 32 cSt and Specific Gravity **SG** = 0.9

- Max Fluid Volume required by the system **Vsmax** = 200 liters.

- Reservoir Material: **Steel**

- Temperature Difference (between reservoir walls and surrounding Air) **ΔT** = 30 $^{\circ}$C.

98

Required (referring to Fig. 2.60): Properly design the reservoir including:
❖ **Reservoir Sizing:**
- Reservoir filling volume of oil **(Vfill)** in liters.
- Oil volume due to thermal expansion **(Vth)** in liters.
- Maximum oil volume in the reservoir **(Vmax)** in liters.
- Air space **(Vas)** in liters.
- Overall Size of the Reservoir **(Vo)** in liters.
- Minimum Absolute Volume **(Vmin)** in liters.

❖ **Cooling:** Determine the reservoir dimensions including:
- Reservoir Length **L** (cm).
- Reservoir Width **W** (cm).
- Area of the reservoir **AR** (cm^2).
- Min oil height **Hmin** (cm).
- Maximum oil height **Hmax** (cm).
- Reservoir height **H** (cm).
- Reservoir Ground Clearance **GCR** (cm).
- Contact area with the oil **AC** (m^2).
- Cooling Capacity **CC** (kW).

99

❖ **Baffle Plate Dimensioning:** Properly design the baffle including:
- Baffle Plate Height **BPH**, Baffle Plate Connecting Area **BPA**, and Width of Connecting Area **BPW**

❖ **Suction Line:**
- Properly size the suction line finding suction line internal diameter **DS**.
- Suction line placement, bottom clearance **SLBC,** and line covered bottom **SLCB.**
- Suction line length **LS** (assuming maximum pressure losses = 0.01barg).
- Check if negative suction pressure **NSP** isn't falling below -3 psig.
- Suction strainer surface area.

❖ **Return Line:**
- Properly size the return line finding return line internal diameter **DR** based on maximum return flow = 200 lit/min.
- Return line bottom clearance **RLBC** and return line covered bottom **RLCB.**

100

100

Solution (referring to Fig. 2.60):

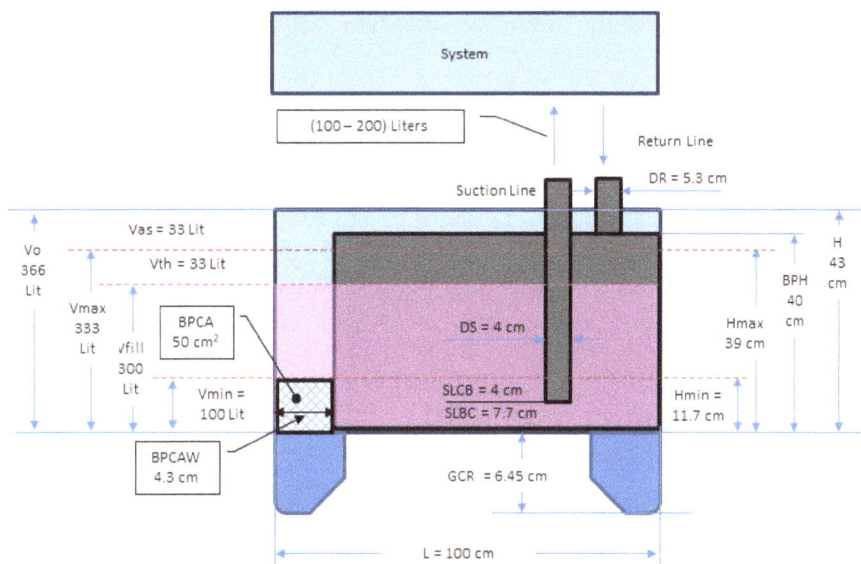

Fig. 2.60- Case study for Hydraulic Reservoir Design   101

❖ Reservoir Sizing:
- Eq. 2.1 → Reservoir filling volume of oil **(Vfill)** = 3 x Qp = 3 x 100 = 300 liters.

- Eq. 2.2 → Oil volume due to thermal expansion **(Vth)** = 0.1 Vfill = 0.1 x 300 = 33 liters

- Eq. 2.3 → Maximum oil volume in the reservoir **(Vmax)** = Vfill + Vth = 300 + 33 = 333 liters. This is worst condition when the machine stopped.

- Eq. 2.4 → Air space **(Vas)** = 0.1 Vmax = 0.1 x 333 = 33.3 liters.

- Eq. 2.5 → Overall Size of the Reservoir **(Vo)** = Vmax + Vas = 333 + 33.3 = 366.3 liters

- Eq. 2.6 → Absolute Minimum Volume **(Vmin)** = Vfill – Vsmax = 300 – 200 = 100 liters.

- It is to be noted that the reservoir must be dimensioned so that this volume is above the lower point of the suction line by at least 5 cm (2 inches).

102

❖ Cooling:
- Eq. 2.7 → Reservoir Length **L** = 84 cm.
- Then Reservoir Length **L** was selected optionally as **L** = 100 cm.
- Eq. 2.8 → Reservoir Width **W** = 0.85 L = 0.85 x 100 = 85 cm.
- Eq. 2.9 → Footprint area of the reservoir **AR** = 100 x 85 = 8500 cm^2
- Eq. 2.10 → Min oil height **Hmin** = Vmin/AR = (100 x 1000) / 8500 = 11.7 cm.
- Eq. 2.11 → Maximum oil height **Hmax** = Vmax/AR = (333 x 1000) / 8500 = 39 cm.
- Eq. 2.12 → Reservoir height **H** = Vo / AR = = (366 x 1000) / 8500 = 43 cm.
- Eq. 2.13 → Reservoir Ground Clearance **(GCR)** = 0.15 x H = 0.15 x 43 = 6.45 cm.
- Eq. 2.14 → Contact area with the oil **AC** = Hmin x 2 (L + W)
$$= 11.7 \times 2 (100 + 85) /10000 = 0.4329 \text{ m}^2$$
- Eq. 2.15A → Cooling Capacity (kW) = AC (m^2) x ΔT (^{0}C) x K
$$= 0.4329 \times 30 \times 0.0058 = 0.753 \text{ kW} = 0.753 \times 3413 = 2570 \text{ BTU/HR}$$

103

❖ Baffle Plate:

- Eq. 2.16 → Baffle Plate Height **BPH**= Hmax + 1 cm = 39 + 1 = 40 cm.

- Eq. 2.17 → Baffle Plate Connecting Area **BPCA** = Qp (lit/min) / 2 = 100 / 2 = 50 cm^2

- Eq. 2.18 → Width of Connecting Area **BPCAW** = BPCA / Hmin (worst condition) = 50 / 11.7 = 4.3 cm.

104

104

❖ Suction Line Diameter **DS**:

- Solution 1: based on the rule of thumb, assuming flow speed = 1 m/s, Eq 2.19A →

$$Q\left[\frac{\text{lit}}{\text{min}}\right] = \frac{v\,[\text{cm/s}] \times A\left[\text{cm}^2\right] \times 60}{1000} \rightarrow A\left[\text{cm}^2\right] = \frac{1000 \times Q\left[\frac{l}{\text{min}}\right]}{v\,[\text{cm/s}] \times 60}$$

$$\rightarrow A\left[\text{cm}^2\right] = \frac{1000 \times 100}{100\,[\text{cm/s}] \times 60} = 16.66 \rightarrow DS[\text{cm}] = \sqrt{\frac{4 \times A}{\pi}}$$

$$= \sqrt{\frac{4 \times 16.66}{\pi}} = 4.6\ \text{cm}$$

105

105

- **Solution 2:** Minimum inner diameter based on Reynold's No. = 2000, Eq 2.20A →

$$DS[mm] = \frac{21231 \, Q\left[\dfrac{l}{min}\right]}{[Cst] \times R_e} = \frac{21231 \times 100}{32 \times 2000} = 33.17 \, mm = 3.3 \, cm$$

- **Solution 3:** Figure 2.61 shows that HCSC validates the calculation resulted from Eq. 2.20A

- **Final Decision** after checking the pump suction port: Suction line diameter is decided to be **DS = 1.5 in = 3.81 cm**.

106

106

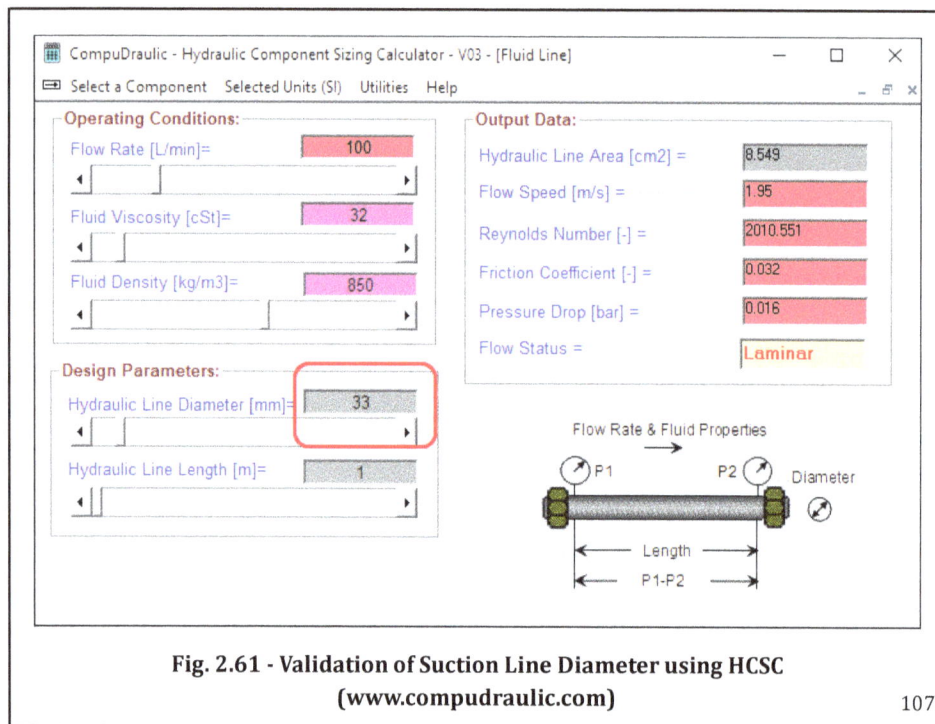

Fig. 2.61 - Validation of Suction Line Diameter using HCSC
(www.compudraulic.com)

107

107

❖ <u>Suction Line Placement, Bottom Clearance, and Covered Bottom:</u>
- Since a baffle plate is used, suction and return lines are placed one on each side of the baffle plate and on the same side of the reservoir.

- Eq. 2.21 → Suction Line Bottom Clearance
- **(SLBC)** = 2 **DS** = 2 x 3.81 ≈ 7.7 cm.

- Eq. 2.22 → Suction Line Covered Bottom
- **SLCB = Hmin – SLBC** = 11.7 – 7.7 = 4 cm.

- Since **SLCB** < 5 cm, then minimum absolute volume of oil **Vmin** can be increased a bit to make the covered bottom **SLCB** equals at least 5 cm.

108

108

❖ <u>Suction Line Length **LS**:</u>
- **First:** Eq. 2.23 → **LS** = 10 **DS** = 10 x 3.81 = 381 mm.

- **Second:** Suction line minimum length is at least 25 cm (10 inches) from last turbulent point → **LS** >= 250 mm from the strainer.

- Then, so far, **LS** >= 381 mm because that satisfies both rules.

- **Third:** Figure 2.62 shows that the suction line length that develops maximum pressure losses of 0.01 bar can be elongated to 1.1 meter.

- Then, finally **381 mm >= LS <= 1100 mm. Decision made to use LS = 500 mm.**

109

109

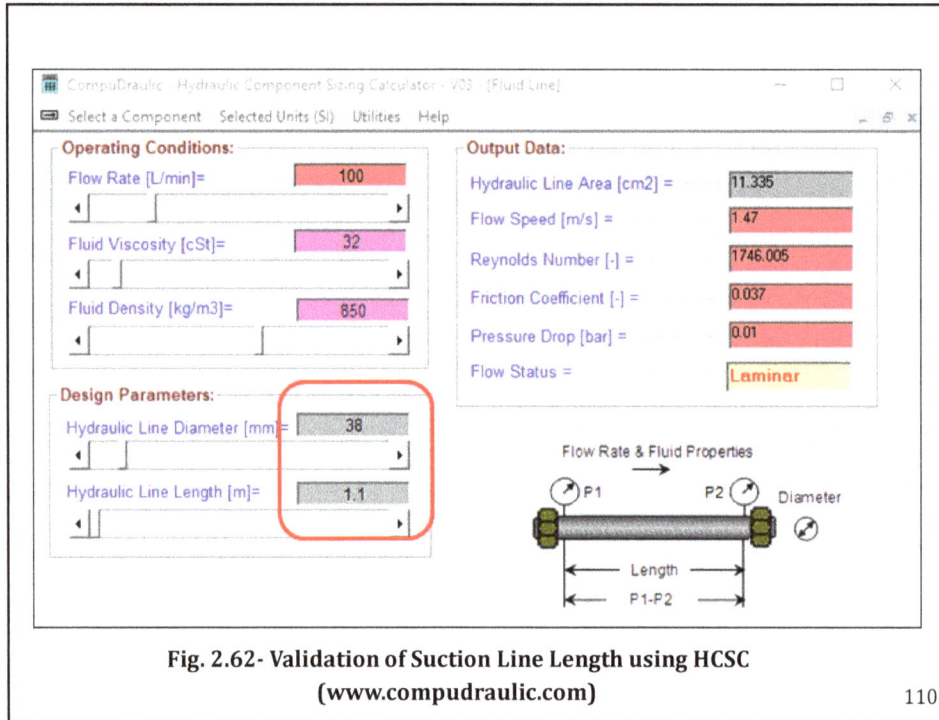

Fig. 2.62- Validation of Suction Line Length using HCSC
(www.compudraulic.com)

110

110

❖ <u>Suction Pressure:</u>

▪ Eq. 2.24A (Hs = Ls = 500 mm & SG = 0.9) →

$$\textbf{NSP (bar)} = -\frac{\textbf{SG} \times \textbf{9.81} \times \textbf{Hs (m)}}{\textbf{100}} = -\frac{\textbf{0.9} \times \textbf{9.81} \times \textbf{0.5}}{\textbf{100}}$$
$$= -\textbf{0.044 barg} = -\textbf{0.6 psig}$$

▪ Negative Suction Pressure (NSP) = -0.6 psig > -3psig, then it meets the requirement.

❖ <u>Suction Strainer Surface Area:</u>

▪ Eq. 2.25A → Suction Strainer Surface Area = 3 x Qp (lit/min) = 3 x 100 = 300 cm².

111

❖ Return Line Diameter **DR**:

▪ Solution 1: A flow distribution has been conducted. Because a differential cylinder is used that has area ratio 2:1, it was found that maximum return flow is 200 lit/min. Therefore, return line size based on the rule of thumb assuming flow speed = 1.5 m/s, Eq 2.19A →

$$Q\left[\frac{lit}{min}\right] = \frac{v\,[cm/s] \times A[cm^2] \times 60}{1000} \rightarrow A[cm^2] = \frac{1000 \times Q\left[\frac{l}{min}\right]}{v\,[cm/s] \times 60}$$

$$\rightarrow A[cm^2] = \frac{1000 \times 200}{150\,[cm/s] \times 60} = 22.21 \rightarrow DR[cm] = \sqrt{\frac{4 \times A}{\pi}}$$

$$= \sqrt{\frac{4 \times 22.21}{\pi}} = 5.3\ cm$$

112

112

▪ Solution 2: Minimum inner diameter based on Reynold's No. = 2000, Eq 2.20A →

$$DR[mm] = \frac{21231\,Q\left[\frac{l}{min}\right]}{v\,[Cst] \times R_e} = \frac{21231 \times 200}{32 \times 2000} = 66.34\ mm = 6.4\ cm$$

▪ Solution 3: Figure 2.63 shows that HCSC validates the calculation resulted from Eq. 2.20A

▪ Final Decision: **DR = 7 cm** = 70 mm.

113

113

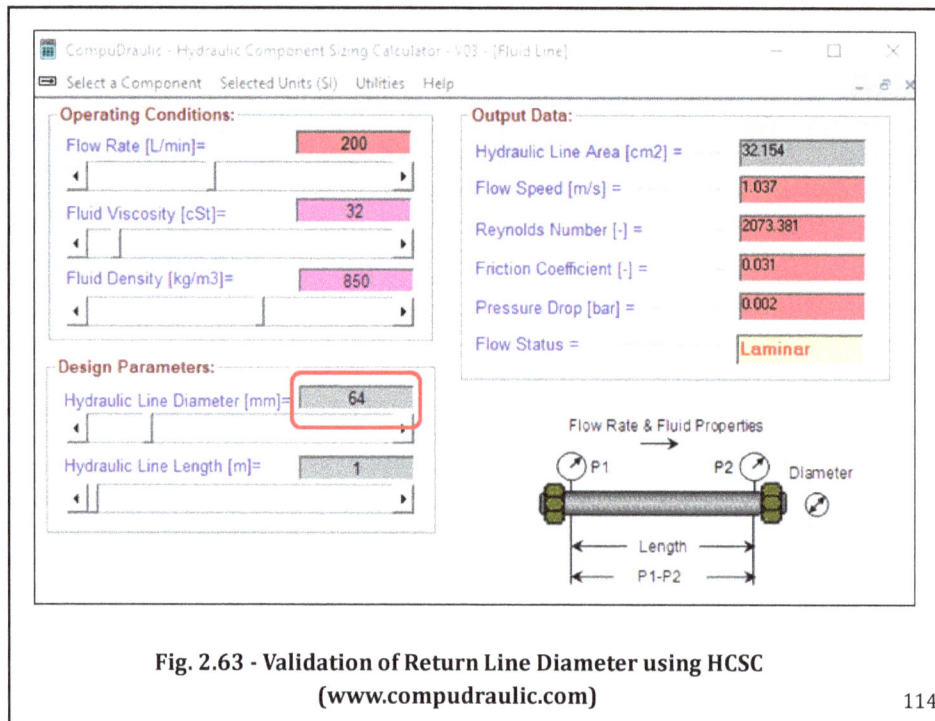

Fig. 2.63 - Validation of Return Line Diameter using HCSC
(www.compudraulic.com)

114

114

- ❖ <u>Return Bottom Clearance, and Covered Bottom:</u>

- ▪ Eq. 2.21 → Return Line Bottom Clearance **(RLBC)** = 2 **DS** = 2 x 7 = 14 cm.

- ▪ Applying Eq. 2.22 → **RLCB = Hmin – RLBC**= 11.7 – 14 = - 2.3 cm, i.e. above **Hmin**!

- ▪ Since **RLBC** > **Hmin**, then the choices here are as follows:

- ▪ <u>Choice 1:</u> Minimum absolute volume of oil **Vmin** can be increased a bit to make the covered bottom **RLCB** equals at least 5 cm.

- ▪ <u>Choice 2:</u> Reducing the bottom clearance to be 5 cm with the consideration that this may result in some back pressure.

115

115

- Choice 3: Repeat reservoir sizing in this backward sequence:
  - Eq. 2.22 → **Hmin** = RLCB + RLBC = 5 cm (recommended) + 14 (calculated) = 19 cm
  - Eq. 2.10 → **Vmin** = Hmin x AR = 19 x 8500 = 161500 cm^3 = 161.5 liters
  - Eq. 2.6 → Vfill = Vmin + Vsmax = 161.5 + 200 = 361.1 Liter.
  - Eq. 2.2 → Oil volume due to thermal expansion **(Vth)**
  - Eq. 2.3 → Maximum oil volume in the reservoir **(Vmax)**
  - Eq. 2.4 → Air space **(Vas)**
  - Eq. 2.5 → Overall Size of the Reservoir **(Vo)**
  - Eq. 2.7 → Reservoir Length **L** = 84 cm.
  - Eq. 2.8 → Reservoir Width **W**
  - Eq. 2.9 → Footprint area of the reservoir **AR**
  - Eq. 2.10 → Min oil height **Hmin**
  - Eq. 2.11 → Maximum oil height **Hmax**
  - Eq. 2.12 → Reservoir height **H**
  - Eq. 2.13 → Reservoir Ground Clearance
  - Eq. 2.14 → Contact area with the oil **AC**
  - Eq. 2.15A → Cooling Capacity (kW)
  - Eq. 2.16 → Baffle Plate Height **BPH**
  - Eq. 2.17 → Baffle Plate Connecting Area **BPCA**
  - Eq. 2.18 → Width of Connecting Area **BPCAW**

116

Chapter 2 Reviews

1. A closed and pressurized tank (as shown in the picture) is commonly used for?
    A. Harshly contaminated work environment.
    B. Aerospace or underwater applications.
    C. Some mobile applications.
    D. All the above.

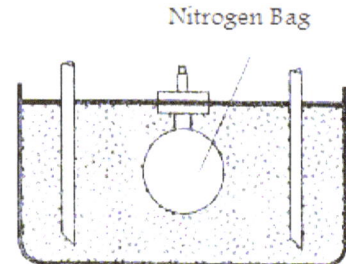

    Nitrogen Bag

2. One of the auxiliary duties of a hydraulic reservoir is?
    A. Maintain oil viscosity.
    B. Air removal from the return oil.
    C. Fast circulation of hydraulic fluid.
    D. Collect leakage from hydraulic components.

3. Suction line is sized primarily to:
    A. Have flow speed does not exceed 5 m/s.
    B. Minimize the losses in the line in order to avoid pump cavitation.
    C. Match the pump intake port.
    D. Only statements B & C are correct.

4. A line connecting a main reservoir and an add-on reservoir is sized to have 1 squared inch for every:
    A. 1 gpm of the pump flow
    B. 2 gpm of the pump flow
    C. 3 gpm of the pump flow
    D. 4 gpm of the pump flow.

5. Air space on top of oil surface in an open reservoir is approximately equal:
    A. 10% of the oil volume
    B. 20% of the oil volume
    C. 30% of the oil volume
    D. 40% of the oil volume

6. Absolute minimum oil volume in an open reservoir should cover the lowest part of the suction line by at least:
    A. 1 inch
    B. 2 inches
    C. 3 inches
    D. 4 inches

7. A reservoir ground clearance is in the order of :
    A. 10% of the reservoir height
    B. 15% of the reservoir height
    C. 20% of the reservoir height
    D. 25% of the reservoir height

8. The baffle plate connecting area is sized based on:
    A. 1 squared cm for every 1 lit/min of the pump flow.
    B. 2 squared cm for every 1 lit/min of the pump flow.
    C. 3 squared cm for every 1 lit/min of the pump flow.
    D. 4 squared cm for every 1 lit/min of the pump flow.

9. Suction line bottom clearance should be in the order of:
    A. (1-2) times the suction line diameter.
    B. (2-3) times the suction line diameter.
    C. (3-4) times the suction line diameter.
    D. (4-5) times the suction line diameter.

10. Suction line length should be in the order of:
    A. 5 times the suction line diameter.
    B. 10 times the suction line diameter.
    C. 15 times the suction line diameter.
    D. 20 times the suction line diameter.

Chapter 2 Assignment

Student Name: --- Student ID: ------------------

Date: -- Score: ------------------------

Assignment:

Calculate the Negative Suction Pressure for a pump that are placed on top of a reservoir with a 1 m suction head. The reservoir contains fluid with specific gravity = 0.9

Solution:

Eq. 2.24A →

$$\text{NSP (bar)} = -\frac{\text{SG} \times 9.81 \times \text{Hs (m)}}{100} = -\frac{0.9 \times 9.81 \times 0.5}{100} = -0.088 \text{ barg}$$
$$= -1.2 \text{ psig}$$

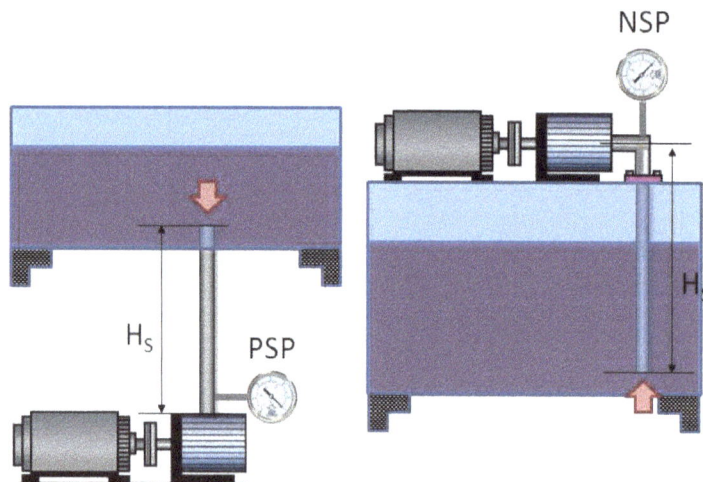

Chapter 3
Hydraulic Transmission Lines

Objectives:

This chapter focuses on browsing the construction and the features of the three main hydraulic transmission lines, Pipes, Tubes, and Hoses. For each transmission line, the following topics are presented: sizing, material, construction and pressure rating. This chapter also presents information about fittings and manifolds.

0

0

Brief Contents:

3.1. Basic Types and Contribution of Hydraulic Transmission Lines

3.2. Sizing of Hydraulic Transmission Lines

3.3- Rated Pressures for Hydraulic Lines

3.4- Hydraulic Pipes

3.5- Hydraulic Tubes

3.6- Hydraulic Hoses

3.7- Flanges for Transmission Line Connections

3.8- Rubber Expansion Fittings

3.9- Test Points

3.10- Pressure Measurement Hoses

3.11 Manifolds

1

1

The following topics are discussed in Chapter 10 in Volume 5
"Safety and Maintenance"

- Transmission Lines Selection and Replacement.

- Transmission Lines Maintenance Scheduling.

- Transmission Lines Installation and Maintenance.

- Transmission Lines Standard Tests and Calibration.

- Transmission Lines Transportation and Storage.

The following topics are discussed in Chapter 10 in Volume 6
"Troubleshooting and Failure Analysis"

- Hydraulic Transmission Lines Inspection.

- Hydraulic Transmission Lines Troubleshooting.

- Hydraulic Transmission Lines Failure Analysis.

2

2

3.1. Basic Types and Contribution of Hydraulic Transmission Lines

- **Purpose:** transmit the energy between system components.

- **Lines vs. Manifolds:**

    Capacitance, Energy Losses, Compactness, External Leakage

- **Types:** Hard (Pipes and Tubes) and Flexible hoses.

Pipes   Tubes   Flexible Hoses

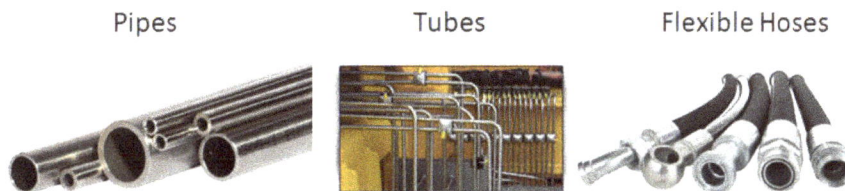

Fig. 3.1 - Types of Hydraulic Transmission Lines

3

3

- **Interchangeability:**
  - Limitedly interchangeable.
  - Each has Pros and Cons for a specific applications.

- **Flexibility:**
  - **Pipes:** Bend (NO) & Flexibility (NO).
  - **Tubes:** Bend (Yes) & Flexibility (NO).   Video 239 (0.5 min)
  - **Hoses:** Bend (Yes) & Flexibility (Yes).
  - **Hoses:** Stretching under pressure + limited working temperature range.

- **Stiffness:**
  - **Pipes and Tubes:** Wall elasticity $\downarrow$.
  - **Hoses (Relatively):** Wall elasticity $\uparrow \rightarrow$ equivalent $\beta \downarrow \rightarrow$
  - system stiffness $\downarrow \rightarrow$ system response & natural frequency $\downarrow$.

- **Service Life:**
  - **Pipes and Tubes:** Service life$\uparrow$.
  - **Hoses:** Service life $\downarrow \rightarrow$ fail without warning $\rightarrow$ routine replacement.

4

4

- **Cost:**
  - **Pipes:** Most cost effective.
  - **Hoses:** Most expensive.

- **Applications:**
  - **Pipes:** Less cost $\rightarrow$ recommended for Industrial applications.
  - **Hoses:** Flexibility $\rightarrow$ recommended for mobile applications.

- **Noise and Vibration:**
  - **Hoses:** Flexibility $\rightarrow$ absorb vibration absorption and noise suppression.

- **Heat Transfer:**
  - **Pipes and Tubes** $\rightarrow$ better heat dissipation than hoses.

5

5

3.2. Sizing of Hydraulic Transmission Lines

❖ Improper (sizing, selection, installation, and maintaining) of hydraulic transmission lines → one or more of the following problems:

- Pump cavitation.
- Turbulent flow.
- System heating-up.
- External leakage.
- Line accidental breakage → possible system damage and loss of life.
- Costs: parts, labor, machine downtime, and cleanup costs.

❖ Sizing a transmission line → calculation of the Inner Diameter (ID).

❖ Sizing methods (mathematically, charts, software, or tables)

6

6

❖ **Sizing transmission lines mathematically Considering:**

- Comply with the recommended flow speed:
- o Equation 2.19 (in Chapter 2), consider the following flow speeds:
- o Pressure Line (mobile machines) = 25 fps.
- o Pressure Line (industrial machines) = 7 - 15 fps. = 2.1- 4.6 m/s.
- o Return/Drain Line = 4 - 7 fps. = 1.2 - 2.1 m/s.
- o Suction Line = 2 - 4 fps. = 0.6 - 1.2 m/s.

- Securing laminar flow:
- o Equation 2.20 (in Chapter 2) is used to size a line based on securing laminar flow in the line (Reynolds Number <2000).

- Considered Results:
- o The larger size out of the two equations is considered.

7

7

❖ Sizing Hydraulic Lines Using Charts:

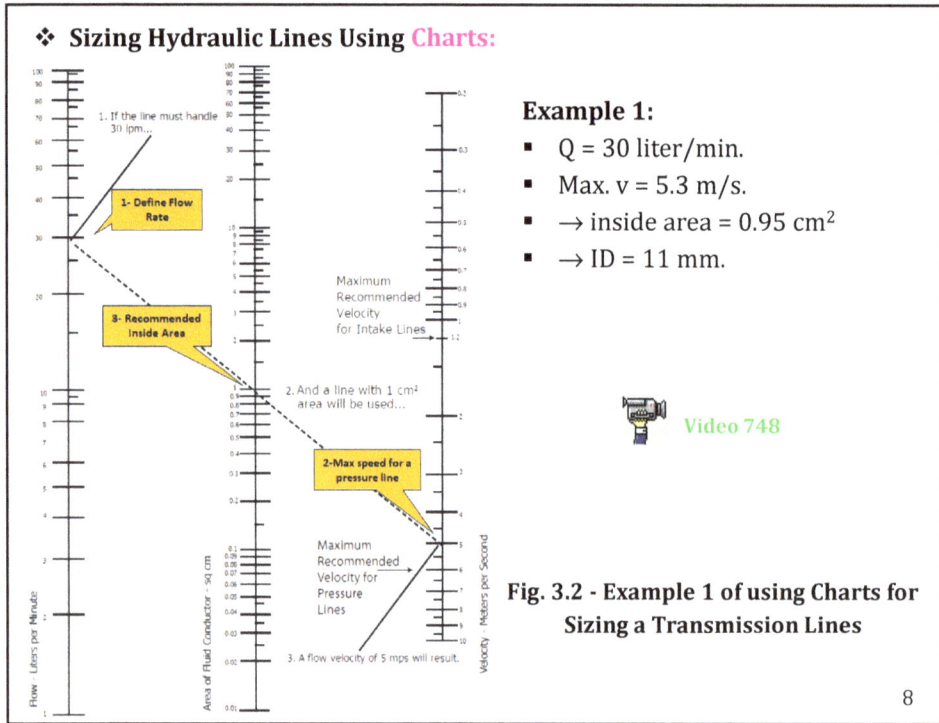

Example 1:

- Q = 30 liter/min.
- Max. v = 5.3 m/s.
- → inside area = 0.95 cm^2
- → ID = 11 mm.

Video 748

Fig. 3.2 - Example 1 of using Charts for Sizing a Transmission Lines

8

8

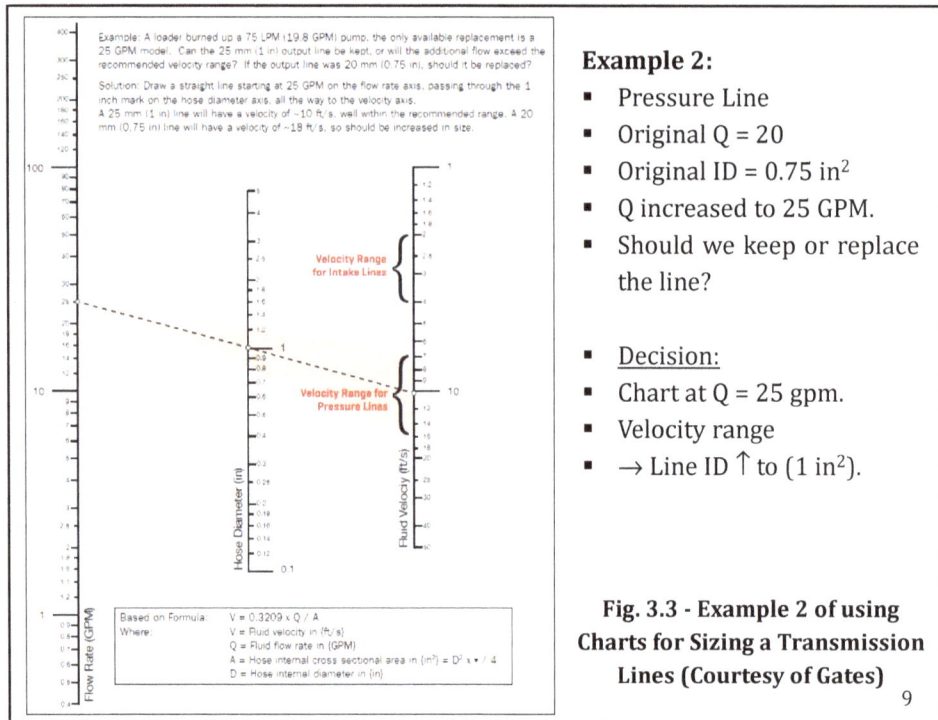

Example 2:

- Pressure Line
- Original Q = 20
- Original ID = 0.75 in^2
- Q increased to 25 GPM.
- Should we keep or replace the line?

- Decision:
- Chart at Q = 25 gpm.
- Velocity range
- → Line ID ↑ to (1 in^2).

Fig. 3.3 - Example 2 of using Charts for Sizing a Transmission Lines (Courtesy of Gates)

9

9

❖ **Sizing Hydraulic Lines Using** Software:
▪ Validation of Example 1

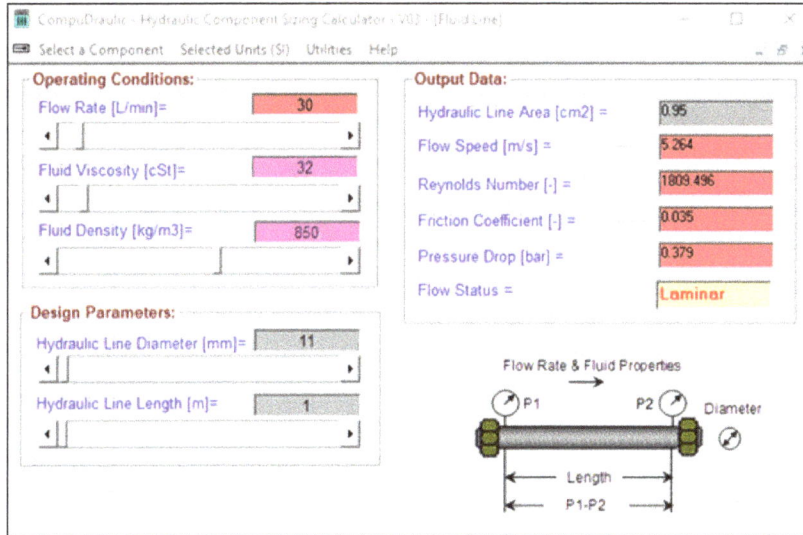

Fig. 3.4 - Using Software for Sizing a Transmission Lines

10

▪ Validation of Example 2
▪ Increasing Q to 25 gpm & keeping ID = 0.75 in2 → Transition flow.

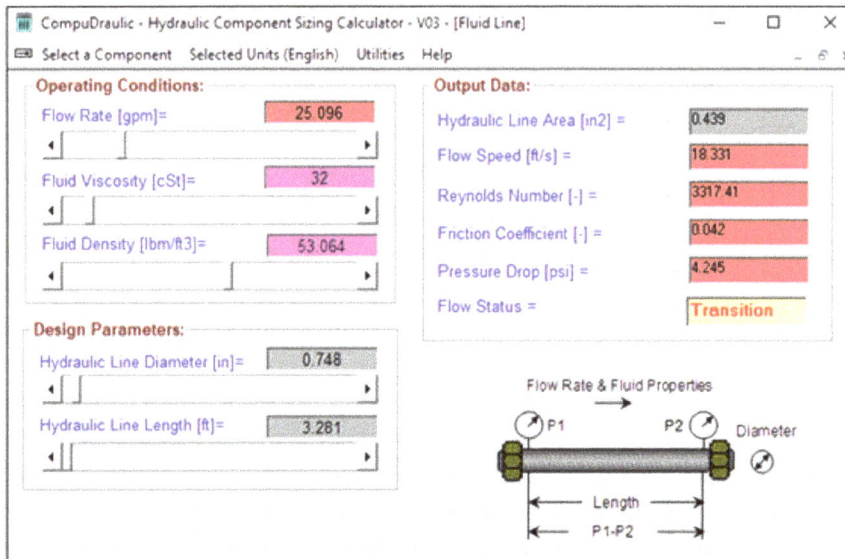

Fig. 3.4 – Continue.

11

❖ **Sizing Hydraulic Lines Using Tables:**

1. Locate flow speed the vertical column that shows with the.

2. Go down in this column until the flow rate through the line is met.

3. Go left to find the corresponding inner diameter.

Example 3:

▪ Flow speed = 15 fps (4.5 m/s) and

▪ Flow rate = 8.9 gpm (40 lit/min)

▪ Inner diameter = 0.493 inches (12.5 mm).

▪ Note: Interpolating the results validates the results from the chart and the software for example 1.

12

12

STANDARD PIPE — SCHEDULE 40

PIPE SIZE	OD (INCHES)	WALL (INCHES)	ID (INCHES)	INT AREA (SQ. INCHES)	WT/FT (POUNDS)	GPM @ 2 FPS	GPM @ 5 FPS	GPM @ 10 FPS	GPM @ 15 FPS	GPM @ 20 FPS	GPM @ 25 FPS
1/8	0.405	0.068	0.269	0.057	0.244	0.4	0.9	1.8	2.7	3.5	4.4
1/4	0.540	0.088	0.364	0.104	0.424	0.6	1.6	3.2	4.9	6.5	8.1
3/8	0.675	0.091	0.493	0.191	0.567	1.2	3.0	6.0	8.9	11.9	14.9
1/2	0.840	0.109	0.622	0.304	0.850	1.9	4.7	9.5	14.2	18.9	23.7
3/4	1.050	0.113	0.824	0.533	1.130	3.3	8.3	16.6	24.9	33.2	41.6
1	1.315	0.133	1.049	0.864	1.677	5.4	13.5	26.9	40.4	53.9	67.4
1 1/4	1.660	0.140	1.380	1.496	2.270	9.3	23.3	46.6	69.9	93.2	116.6
1 1/2	1.900	0.145	1.610	2.036	2.715	12.7	31.7	63.5	95.2	126.9	158.7
2	2.375	0.154	2.067	3.356	3.649	20.9	52.3	104.6	156.9	209.2	261.5
2 1/2	2.875	0.203	2.469	4.788	5.787	29.8	74.6	149.2	223.9	298.5	373.1
3	3.500	0.216	3.068	7.393	7.568	46.1	115.2	230.4	345.7	460.9	576.1
3 1/2	4.000	0.226	3.548	9.887	9.100	61.6	154.1	308.2	462.3	616.4	770.5
4	4.500	0.237	4.026	12.730	10.779	79.4	198.4	396.8	595.2	793.7	992.1
5	5.563	0.258	5.047	20.006	14.602	124.7	311.8	623.6	935.4	1,247.2	1,559.1
6	6.625	0.280	6.065	28.890	18.954	180.1	450.3	900.6	1,350.9	1,801.1	2,251.4
8	8.625	0.322	7.981	50.027	28.524	311.9	779.7	1,559.4	2,339.2	3,118.9	3,898.6
10	10.750	0.365	10.020	78.854	40.441	491.6	1,229.0	2,458.1	3,687.1	4,916.1	6,145.1
12¹	12.750	0.406	11.938	111.932	53.469	697.8	1,744.6	3,489.1	5,233.7	6,978.3	8,722.9
12²	12.750	0.375	12.000	113.097	49.510	705.1	1,762.7	3,525.5	5,288.2	7,051.0	8,813.7

Table 3.1

13

13

3.3- Rated Pressures for Hydraulic Lines

❖ **Burst Pressure:**

- **Definition:** is the maximum static pressure above which the lines assembly fails.
- Hoses → depends on the structure of the hose.
- Hoses → determined by standard test procedure.

- Pipes and Tubes → depends on the material and wall thickness.
- Pipes and Tubes → determined by standard test procedure
- Pipes and Tubes → estimated by Barlow's formula

$$BP = (2 \times t \times s)/OD \qquad 3.1$$

Where:

- BP = Burst Pressure (psi)
- **t** = Wall Thickness [in]
- **s** = Ultimate Strength of Material (psi)
- **OD** = Outer Diameter of Pipe or Tube

14

14

❖ **Maximum Allowable Pressure:**

$$WP = (BP/SF) \qquad 3.2$$

Where:

- WP = Allowable Working Pressure.
- BP = Burst Pressure.
- SF = Safety Factor = typically (4-6).
- BUT may be different in some applications.

15

15

3.4- Hydraulic Pipes
3.4.1- Features of Hydraulic Pipes

- Rigid, not flexible or intend to bend.
- Highest tensile strength.
- Least expensive.
- Transmit large power (flow and pressure) for long distances.
- Largest (weight/unit length).
- Must be clamped to a fixed frame to carry the weight of the pipe.
- Used for both industrial and mobile applications.

Fig. 3.5 - Use of Hydraulic Pipes in Industrial (Left) and Mobile (right) Applications

16

16

3.4.2- Material of Hydraulic Pipes

- Different grades of Carbon Steel depending on the pressure rating.
- known as black iron pipes that may be either hot or cold drawn.
- Hydraulic fluids additives react with the zinc coating →
- Galvanized pipes should never be used in hydraulic systems.

3.4.3- Sizing of Hydraulic Pipes

- **Nominal Size** → are the outer diameter "**OD**" where fitting is assembled on.
- **Pipe Sizes Range** → (0.5 - 12) inches → (10-300) mm.
- **ANSI Standard** → Schedule Number → different pressure ratings.
- **High Pressure** →
  - Walls thickness ↑→ Schedule # ↑ → Weight/Unit length ↑
  - → smaller **ID** → Possible **turbulent flow** & Higher **ΔP**.
  - So, larger **OD** is required to maintain same **ID**, laminar flow & reasonable **ΔP**.

17

17

NOMINAL SIZE	PIPE O.D.	SCHED. 40	SCHED. 80	SCHED. 160	DOUBLE EXTRA HEAVY
1/8	.405	.269	.215	--	--
1/4	.540	.364	.302	--	--
3/8	.675	.493	.423	--	--
1/2	.840	.622	.546	.466	.252
3/4	1.050	.824	.742	.614	.434
1	1.315	1.049	.957	.815	.599
1 1/4	1.660	1.380	1.278	1.160	.896
1 1/2	1.900	1.610	1.500	1.338	1.100
2	2.375	2.067	1.939	1.689	1.503
2 1/2	2.875	2.469	2.323	2.125	1.771
3	3.500	3.068	2.900	2.624	--
3 1/2	4.000	3.548	3.364	--	--
4	4.500	4.026	3.826	3.438	--
5	5.563	5.047	4.813	4.313	4.063

INSIDE DIAMETER

Schedule 40

Schedule 80

Schedule 160

Double Extra Heavy

Table 3.2- ANSI Standard Schedules for Hydraulic Pipes

18

18

Video 159 (1.5 min)

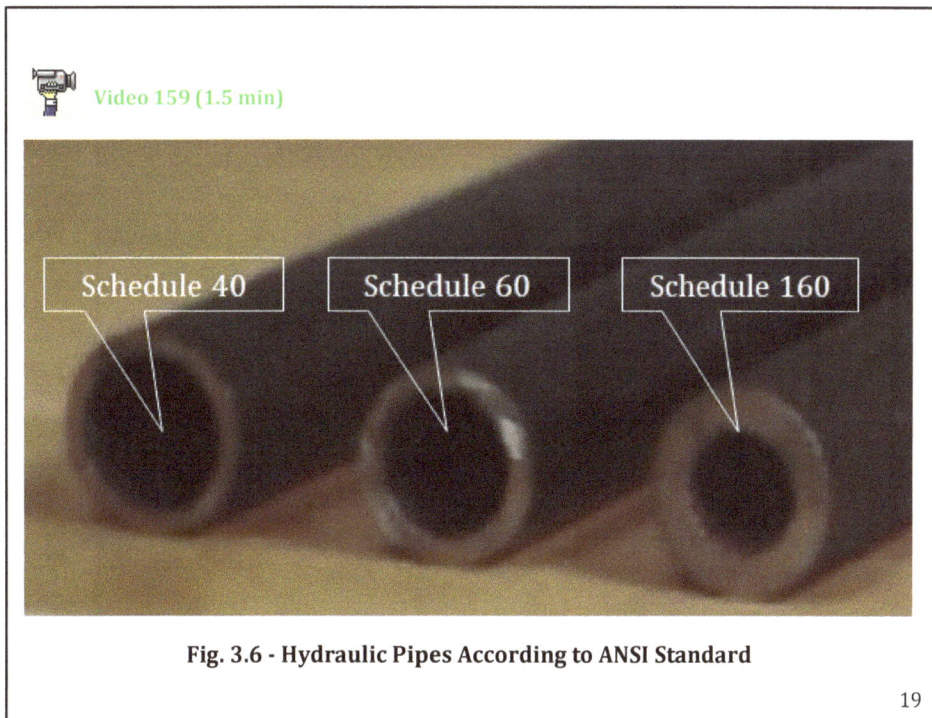

Fig. 3.6 - Hydraulic Pipes According to ANSI Standard

19

19

Example of Pipe Size Selection:

Problem Definition: A pump drives a hydraulic motor. Flow distribution analysis is conducted and shows that maximum flow in suction, pressure and return line is 40 GPM.

Required: Referring to Table 2.3 and 3.3, find the nominal and inside diameter of each line. Both the suction and return pipes shall be selected from schedule 40 and pressure pipe from schedule 80.

Suction Line (Purple): Allowable flow speed = 2 FPS, closest upper flow = 46.1 GPM → Pipe Nominal Size = 3 → ID = 3.068 in (Schedule 40).

Return Line (Blue): Allowable flow speed = 5 FPS, closest upper flow = 52.3 GPM → Pipe Nominal Size = 2 → ID = 2.067 in (Schedule 40).

Pressure Line (Red): Allowable flow speed = 10 FPS, flow = 40 GPM → Pipe Nominal Size = 1.25 → ID = 1.278 in (Schedule 80).

20

20

Table 3.3 - Examples of Sizing Hydraulic Pipes

21

21

3.4.4- Pressure Rating of Hydraulic Pipes

- Table 3.4 → Tabulated by ANSI B 31.3 based on:

  o **Burst Pressure** is based on Barlow's formula.

  o For seamless steel pipe material (ASTM A53)

  o Or A106 Grade B steel with tensile strength of 60k psi.

- **Example (a pipe size of nominal size =1):**

  o Estimated burst pressure ≈ 6k psi in schedule 40,

  o Estimated burst pressure ≈ 10k psi in schedule 80,

  o Estimated burst pressure ≈ 16k psi in schedule 160.

  o **Maximum Allowable Working Pressure**

  o Is based on various safety factors:

  o In schedule 40, SF ≈ SF 9.

  o In schedule 80, SF ≈ 6.

  o In schedule 160, SF ≈ 5.

22

22

Threaded Connections Pressure Ratings of Seamless Steel

1 inch OD Different schedule

PIPE SIZE & SCHEDULE	ALLOWABLE WORKING PRESSURE (PSIG)		ESTIMATED BURST VALUE P_B (PSIG)	WATER HAMMER FACTOR
	ANSI B31.1	ANSI B 31.3		
1/8-40	994	746	11,369	327.6
1/8-80	3,509	2,632	19,369	512.9
1/4-40	948	711	9,680	178.9
1/4-80	3,103	2,327	16,560	260.0
3/8-40	916	687	8,277	97.55
3/8-80	2,864	2,148	14,500	132.5
1/2-40	884	663	7,409	61.28
1/2-80	2,576	1,932	12,857	79.59
1/2-160	4,445	3,334	18,551	109.2
1/2-XXS	8,577	6,433	33,837	373.3
3/4-40	842	631	6,384	34.92
3/4-80	2,295	1,722	11,070	43.06
3/4-160	4,704	3,528	18,384	62.89
3/4-XXS	7,328	5,496	28,670	125.9
1-40	829	622	5,788	21.55
1-80	2,126	1,595	9,086	25.89
1-160	4,249	3,187	16,465	35.69
1-XXS	6,804	5,103	26,321	66.08
1 1/4-40	806	605	5,091	12.45
1 1/4-80	1,941	1,456	8,778	14.52
1 1/4-160	3,319	2,490	13,043	17.62
1 1/4-XXS	6,575	4,932	22,585	29.55
1 1/2-40	798	598	4,764	9.147
1 1/2-80	1,866	1,399	8,236	10.54
1 1/2-160	3,522	2,642	13,353	13.24
1 1/2-XXS	6,096	4,572	20,868	19.59
2-40	773	580	4,266	5.549
2-80	1,764	1,323	7,500	6.306
2-160	3,834	2,875	13,806	8.331
2-XXS	5,435	4,076	18,514	10.50
2 1/2-40	815	611	4,299	3.889
2 1/2-80	1,750	1,312	7,348	4.394
2 1/2-160	3,074	2,305	11,478	5.250
2 1/2-XXS	5,588	4,191	18,866	7.559
3-40	801	601	3,977	2.519
3-80	1,683	1,263	6,857	2.819
3-160	3,201	2,401	11,589	3.443
3-XXS	5,096	3,822	17,143	4.482
3 1/2-40	790	592	3,780	1.883
3 1/2-80	1,636	1,226	6,540	2.093
3 1/2-XXS	4,783	3,587	16,080	3.186
4-40	789	592	3,653	1.463
4-80	1,604	1,203	6,320	1.620
4-160	3,263	2,447	11,493	2.006
4-XXS	4,556	3,417	15,307	2.386
5-40	772	579	3,408	0.9308
5-80	1,543	1,157	5,932	1.023
5-160	3,270	2,453	11,326	1.275
5-XXS	4,178	3,134	14,021	1.436
6-40	766	575	3,260	0.6445
6-80	1,608	1,206	6,014	0.7144
6-160	3,275	2,456	11,212	0.8812
6-XXS	4,159	3,120	13,838	0.9887

DISCONTINUED FORMER STANDARD

Table 3.4 - ANSI B 31.3 Standard for Hydraulic Pipes Pressure Rating   23

23

3.4.5- Hydraulic Pipe Assembly
3.4.5.1- Hydraulic Pipes Threads

Video 746

❖ **Pipe Connection:**
- Pipes used in hydraulic systems should be seamless.
- Pipes are traditionally assembled using tapered threads.
- External threads cut on the pipe ends.
- Internal threads cut into the openings of the pipe fittings.
- Tapered thread (3/4" per foot) → positive seal when properly tightened.

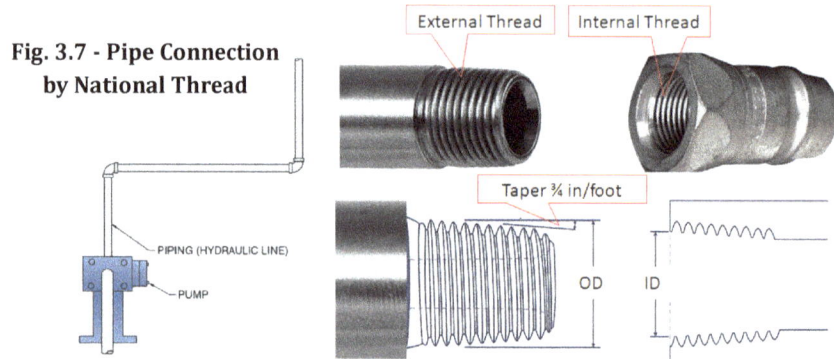

Fig. 3.7 - Pipe Connection by National Thread

External Thread   Internal Thread

Taper ¾ in/foot

PIPING (HYDRAULIC LINE)

PUMP

OD   ID

24

24

❖ **Spiral vs. Dryseal Thread:**
- Standard (Spiral) National Pipe Tapered (NPT):
  - Even when tightened → spiral clearance →
  - Tends to leak under high pressure.

- Dryseal National Pipe Tapered (NPTF) thread:
- It is recommended for high pressure and fuel lines.

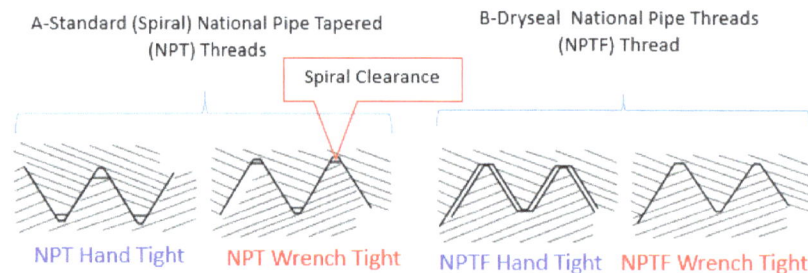

A-Standard (Spiral) National Pipe Tapered (NPT) Threads

B-Dryseal National Pipe Threads (NPTF) Thread

Spiral Clearance

NPT Hand Tight   NPT Wrench Tight   NPTF Hand Tight   NPTF Wrench Tight

Fig. 3.8 - Standard Spiral Thread versus Dryseal Thread

25

25

3.4.5.2- Fittings for Hydraulic Pipes

- A wide variety of threaded connections are available to connect a pipe to hydraulic system.

- Alternatively, and more recommended.

  (Butt & Socket Weld fittings and 4 Bolts 61/62 SAE Flanges)

  Video 162 (0.5 min)

26

26

Fig. 3.9-
Hydraulic
Pipe Fittings

27

27

3.5- Hydraulic Tubes

3.5.1- Features of Hydraulic Tubes (as compared to pipes)

- Are semirigid and can be bent (with the minimum bend radius respected).
- Have relatively thinner walls and less weight per unit length.
- Are easier to assemble (no welding is needed).
- Are connected to the components by various types of fittings and flanges.
- Have lower pressure ratings (as compared to Pipes).
- Used for both industrial and mobile applications.

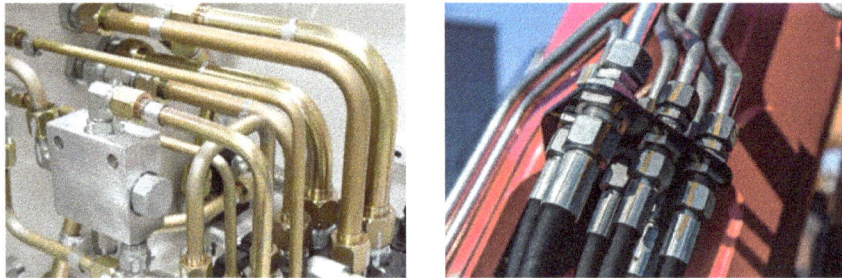

**Fig. 3.10 - Use of Hydraulic Tubes
in Industrial (Left) and Mobile (right) Applications**

Video 160 (0.5 min)

28

28

3.5.2- Material of Hydraulic Tubes

Hydraulic tubes are produced from various material as follows:

- Stainless Steel: Ultimate Strength = 60/75/100 thousands of psi
- Low Carbon Steel: Ultimate Strength = 55,000 psi
- Cooper: Ultimate Strength = 32,000 psi

Stainless Steel
Tubes

Copper Tubes

Fig. 3.11 - Material of Hydraulic Tubes

29

29

3.5.3- Size of Hydraulic Tubes

- Like hydraulic pipes, nominal size of hydraulic tubes is "OD".
- Tube sizes are ranging between (0.125 – 2.25) inches (4-42 mm).

3.5.4- Pressure Rating of Hydraulic Tubes

- Mathematics for estimating burst pressure:
- Barlow's formula "1", Boardman's Formula "2", and Lame's Formula "3".

- Barlow's formula → lowest burst pressure among the three formula
- → only Barlow's formula is considered in this textbook (like pipes)

- Table 3.5 → Maximum Allowable Pressure is Tabulated based on:
  o Carbon Steel (Ultimate Strength = 55000 psi)
  o Safety factor = 4:1.

30

Nominal Tube OD, in.		See Note*	Nominal Tube Wall Thickness, in.									
			0.028	0.035	0.049	0.065	0.083	0.095	0.109	0.120	0.134	0.148
1/8	0.125	1	5,800	7,000								
		2	6,800	9,000								
		3	6,650	8,450								
3/16	0.188	1	3,750	4,650								
		2	4,250	5,500								
		3	4,250	5,450								
1/4	0.250	1	2,800	3,500	4,900	6,500						
		2	3,100	3,950	5,800	8,200						
		3	3,100	3,950	5,750	7,800						
5/16	0.312	1	2,250	2,800	3,900	5,200						
		2	2,400	3,100	4,500	6,250						
		3	2,450	3,100	4,500	6,150						
3/8	0.375	1	1,850	2,350	3,250	4,350	5,550	6,350				
		2	2,000	2,500	3,650	5,050	6,700	7,950				
		3	2,000	2,550	3,650	5,000	6,550	7,600				
1/2	0.500	1		1,750	2,450	3,250	4,150	4,750	5,450	6,000		
		2		1,850	2,650	3,650	4,800	5,600	6,600	7,450		
		3		1,850	2,700	3,650	4,800	5,550	6,450	7,200		
5/8	0.625	1		1,400	1,950	2,600	3,300	3,800	4,350	4,800		
		2		1,450	2,100	2,850	3,700	4,350	5,050	5,650		
		3		1,500	2,100	2,850	3,750	4,350	5,050	5,600		
3/4	0.750	1		1,150	1,650	2,150	2,750	3,150	3,650	4,000		
		2		1,200	1,700	2,350	3,050	3,500	4,100	4,600		
		3		1,200	1,750	2,350	3,050	3,550	4,150	4,600		
7/8	0.875	1		1,000	1,400	1,850	2,350	2,700	3,100	3,400		
		2		1,050	1,450	1,950	2,550	2,950	3,450	3,850		
		3		1,050	1,500	2,000	2,600	3,000	3,500	3,900		
1	1.000	1		875	1,200	1,600	2,050	2,350	2,700	3,000	3,350	3,700
		2		900	1,250	1,700	2,200	2,550	3,000	3,300	3,750	4,200
		3		900	1,300	1,750	2,250	2,600	3,000	3,350	3,800	4,200

Note: Wall thicknesses having values shown to the right of the bold line are not normally considered suitable for 37° single flaring to J533.

Table 3.5 – Maximum Allowable Working Pressure (psi) for Carbon Steel Tubes

31

3.5.5- Hydraulic Tubes Assembly
3.5.5.1- Tube Bending

Unlike Pipes, Tubes can be Bent

- Tube bending → outer surface wall thickness ↓ → pressure rating ↓ →
- Minimum bend radius should be respected.
- Minimum bend radius is reported by tube manufacturer.

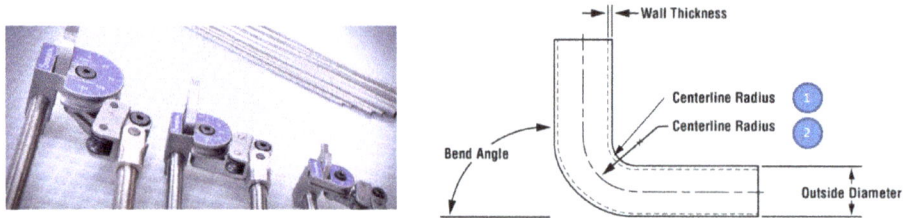

Fig. 3.12 - Manual Mandrills of Different Sizes for Tube Bending

32

Note: For large and small ODs, closed bends aren't allowed.

TUBE OD (INCHES)	CLOSE① BEND (INCHES)	NORMAL BEND (INCHES)	MS33611 STANDARD	METRIC SIZES		
				TUBE OD (MM)	CLOSE① BEND (MM)	NORMAL BEND (MM)
0.125	—	0.281①	—	4	—	9
0.188	—	0.422①	—	6	—	13
0.250	0.562	0.750	—	8	—	18
0.312	0.687	1.000	—	10	32	32
0.375	0.937	1.250	1.125	12	—	32
0.500	1.250	2.000	1.500	15	38	38
0.625	1.500	2.500	1.875	16	38	38
0.750	1.750	3.000	2.250	18	44	44
0.875	2.000	3.500	2.625	20	44	44
1.000	3.000	4.000	3.000	22	—	89
1.125	3.500	4.500	3.375	25	—	100
1.250	3.750	5.000	3.750	28	—	112
1.500	5.000	6.000	4.500	30	—	128
1.750	—	7.000	—	35	—	140
2.000	—	8.000	—	38	—	152
2.250	—	9.000①	—	42	—	168

Table 3.6 - Minimum Bend Radius for Hydraulic Tubes of Different Sizes

33

3.5.5.2- Hydraulic Tubes Flaring

Unlike Pipes, Tubes can be Flared
- **SAE J533B Standard** (37° and 45°) & (Single or Double) Flaring

For more info on tube flaring, review in textbook:

- Fig. 3.13
- Table 3.7
- Fig. 3.14
- Table 3.8

34

34

3.5.5.3- Identification of Fittings for Hydraulic Tubes

❖ **What is a Tube Fitting?**
- A connection between a transmission line and a hydraulic component.
- Available in various sizes, materials, and configurations.

❖ **How to Identify a Fitting?**
- Body Configuration.
- Seal Configuration.
- Standards.
    o Thread Characteristics.
    o Size.
    o Material.
    o Interchangeability.

❖ **Fitting Standards:**
- North American "JIC/SAE/NPT".
- British "B".
- Metric French "GAZ".
- German "DIN".
- Japanese "JIS".
- International standard "ISO".

35

35

❖ **General Body Configurations in Fittings:**

Fixed Joins:
1. Straight Male Stud Fitting
2. Elbow Banjo Fitting
3. Equal Elbow Fitting
4. Equal T Fitting
5. Straight Male Fitting
6. Equal Cross Fitting

Rotating "Swivel" Joints:
1. Seal
2. Ball Race

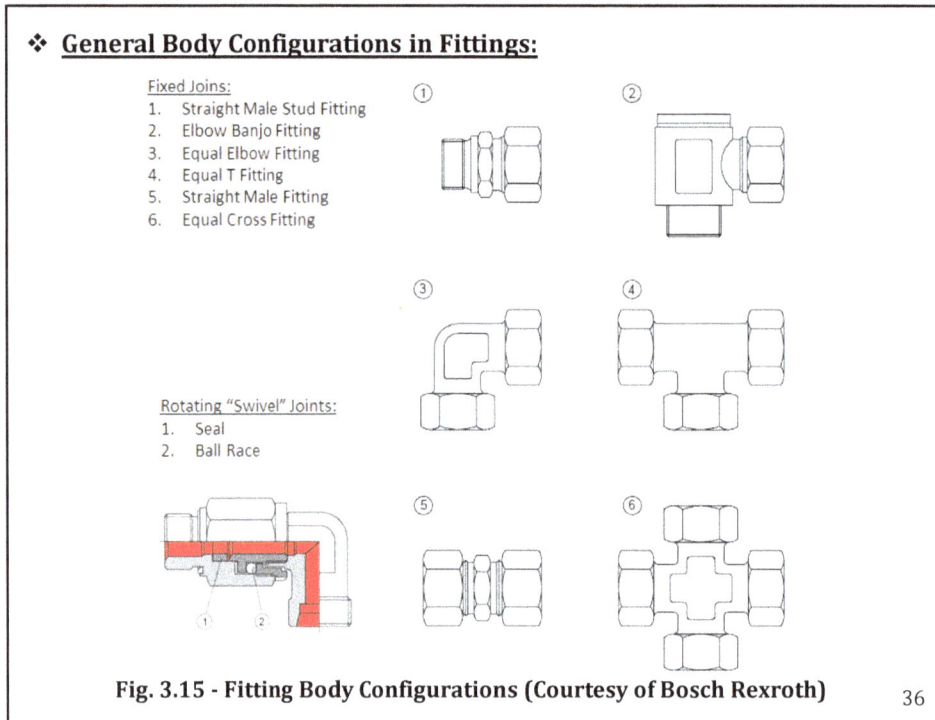

Fig. 3.15 - Fitting Body Configurations (Courtesy of Bosch Rexroth)   36

36

❖ **General Seal Configuration in Fittings:**
- **O-Ring Sealing:**
  - Sealing due to O-ring compression.
  - The nipple is welded to the tube.
  - Excellent for high-pressure applications.
- **Flareless Mated Surfaces (Mechanical Joints) Sealing:**
  - Sealing due to deformation of Compression Ring "Cutting Ring".
- **Flareless Mated Surfaces (Mechanical Joints) Sealing:**
  - Sealing between flared tube & fitting with coned seats (different seat angles).
- **Thread Interference Sealing:**
  - Sealing due to thread deformation when tightened.
  - Is the easiest type of fitting to use.

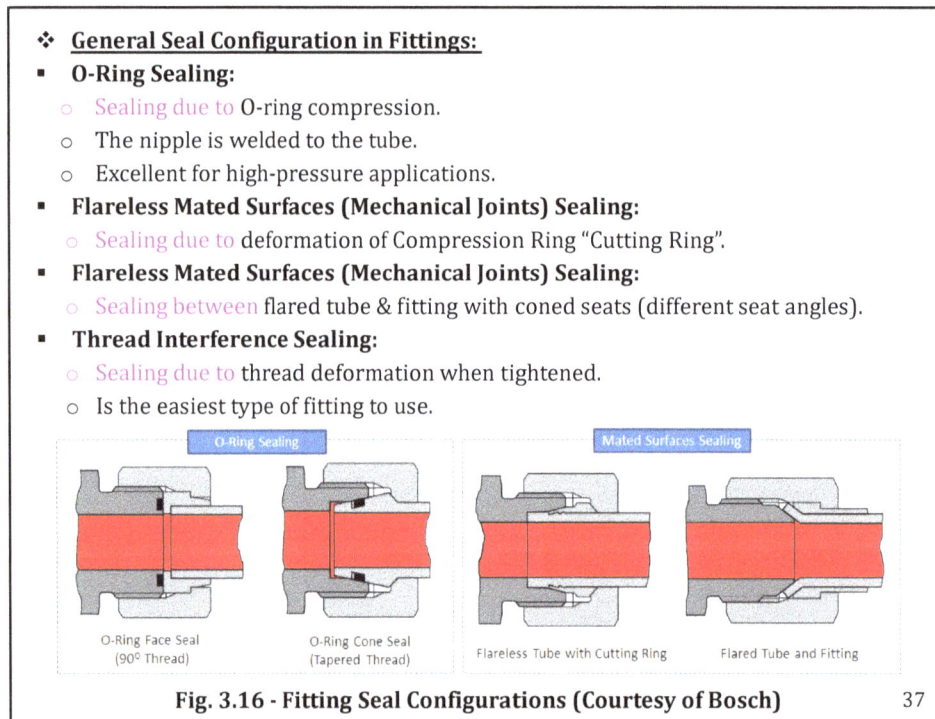

Fig. 3.16 - Fitting Seal Configurations (Courtesy of Bosch)   37

3.5.5.4- Threaded Sealing Fittings

❖ **North American Standards Thread Types:**

- Is widely used for over 100 years.
- Used to effectively seal pipes for fluid and gas transfer.
- Are available in iron or brass for low-pressure applications.
- Are available in carbon steel and stainless steel for high-pressure.

- Designations:

N = National, **P** = Pipe, **S** = Straight,

T = Tapered, **F** = Fuel (Dryseal) , and **M** = Mechanical Joints.

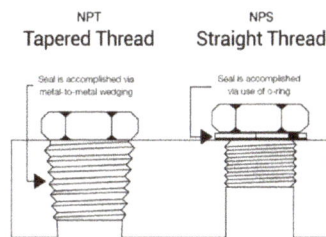

38

38

- **National Pipe Tapered (NPT):**
  o Is a Spiral tapered thread → spiral clearance → low pressure application.
  o Larger fitting Size & frequent use → Spiral clearance ↑.

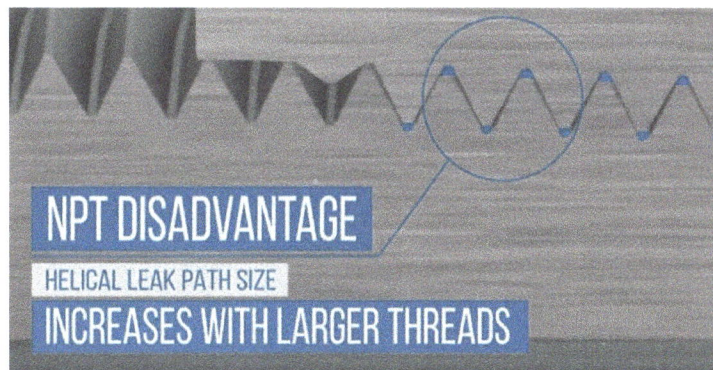

Fig. 3.17 – Spiral Clearance on NPT Fittings
(Courtesy of Brennan Industries)

39

39

- **National Pipe Tapered for Fuels (NPTF):**
  - Dryseal tapered thread for Fuel (made to specifications ASME 1.20.3).
  - It is used for both male and female ends.
  - The seal takes place by deformation of the threads when tightened.

- **National Pipe Straight for Fuels (NPSF):**
  - Dryseal straight thread.

- **National Pipe Straight for Swivel Mechanical Joints (NPSM):**
  - Female end with straight thread for Mechanical Joints.
  - 30° internal chamfer.
  - The seal takes place on the 30° seat.

Fig. 3.18 - National Pipe Thread (Courtesy of Gates)

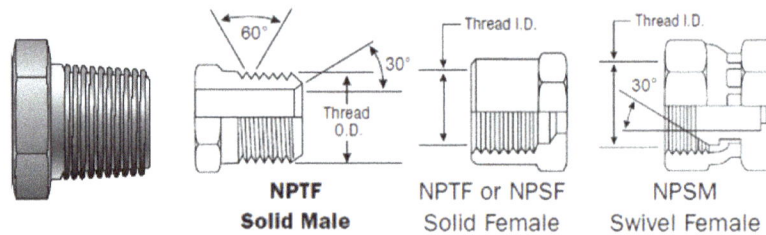

40

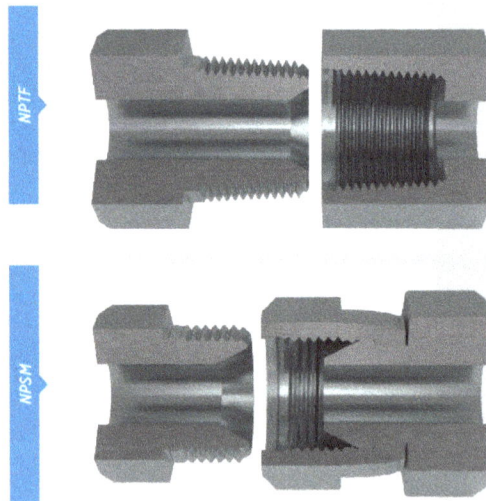

Video 686 (4 min)

**Fig. 3.19 – Crossectional Views of NPTF and NPSM Fittings
(Courtesy of Brennan Industries)**

41

❖ **North American Standards Thread Types (SAE):**

- **Male:** straight threads and SAE 45° inverted flare.

- **Female:** straight threads and SAE 42° inverted flare.

- **Seal:** Between the flare seats on this male and female only

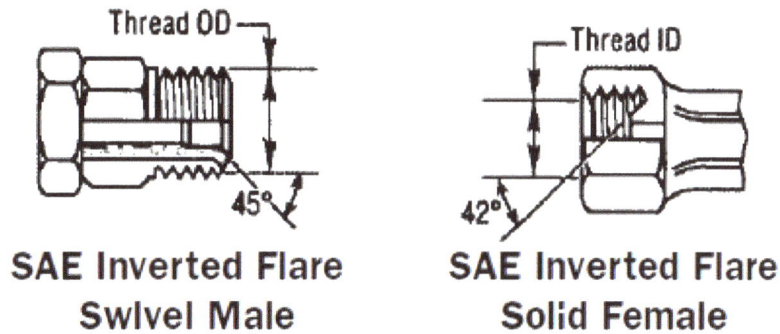

Fig. 3.20 - SAE Inverted Flare Thread (Courtesy of Gates)

42

42

Threaded Sealing Fittings in other Standards, review in textbook:

- **Figures 3.21 through 3.25**
- **Figures 3.21 through 3.30**

43

43

- **Example of DIN 2353 - Innovative Metric Fitting (EO-2):**
  - Description:
  - Is a high-pressure tube fitting generation.
  - Is a metric straight thread design according to

    ISO 8434-1, DIN 2353 or DIN 3861.

  - It consists of a body, a functional nut, and an elastomeric seal.

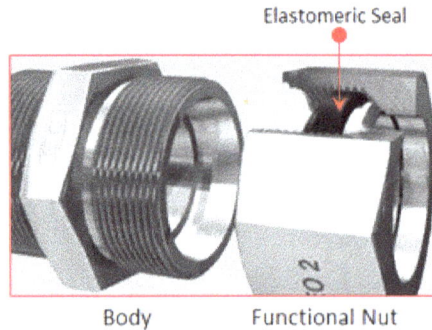

Fig. 3.26 - EO-2 Fitting (Courtesy of Parker)   44

44

- Operation:
- The elastomeric seal
- → compensates for all manufacturing tolerances.
- → located in between the inner cone of the fitting body.
- → assures a hermetically sealed tube joint.
- → eliminates air-ingress into the fluid system.

1- Tightened by Hand

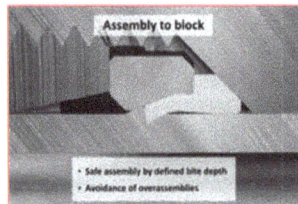

2- Tightened by Wrench
to Specified Torque

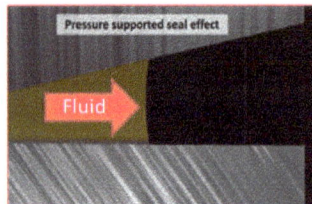

3- Fluid Pressure Applies

Video 652 (1.5 min)

Fig. 3.26 - Continue

45

45

3.5.5.5- O-Ring Sealing Fittings

❖ **North American Standards (SAE):**

- **SAE-JI926 Straight Thread** O-Ring Boss (ORB):
- o **Male**: has an O-ring.
- o **Female:** found on ports.
- o **Thread:** straight.
- o **Seal:** Between the O-ring (male) and the sealing chamfer (female).
- o **Applications:**
  - ✓ Pressure: **Ok** for Medium-High pressure applications.
  - ✓ Temperature: **Limited** use in high-temperature applications.

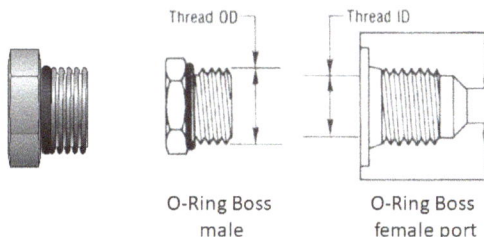

Fig. 3.31 - SAE Straight Thread O-Ring Boss

46

46

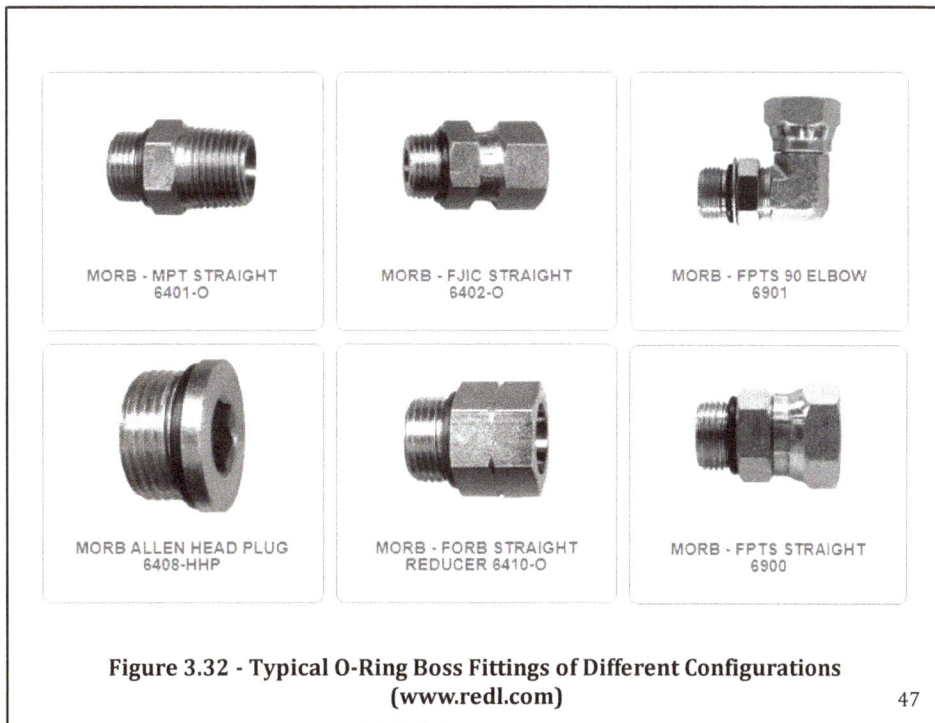

Figure 3.32 - Typical O-Ring Boss Fittings of Different Configurations (www.redl.com)

47

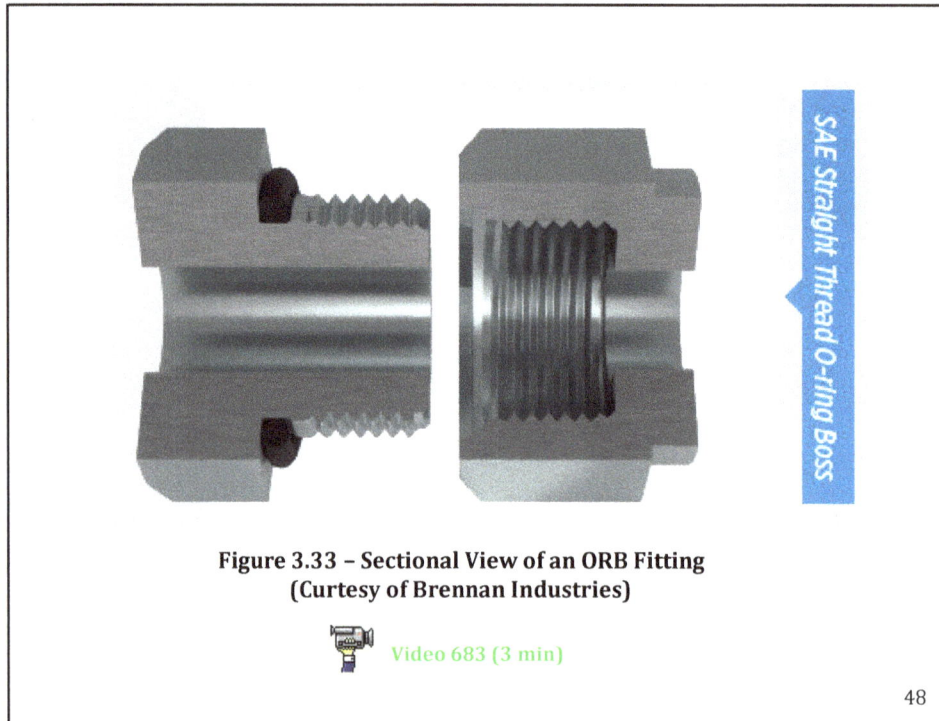

SAE Straight Thread O-ring Boss

Figure 3.33 – Sectional View of an ORB Fitting
(Curtesy of Brennan Industries)

Video 683 (3 min)

48

48

- **O-Ring Face Seal (ORFS) - SAE -J1453:**
  o **Male:** O-Ring is placed in a captive groove on the face of the fitting.
  o **Female:** Swivel female (SAE J1453 fitting only).
  o **Seal:** Between the O-ring (male) and the flat face (female).
  o **Applications:**
    ✓ Pressure: **Ok** for Medium-High pressure applications.
    ✓ Temperature: **Limited** use in high-temperature applications.

Video 682 (3 min)

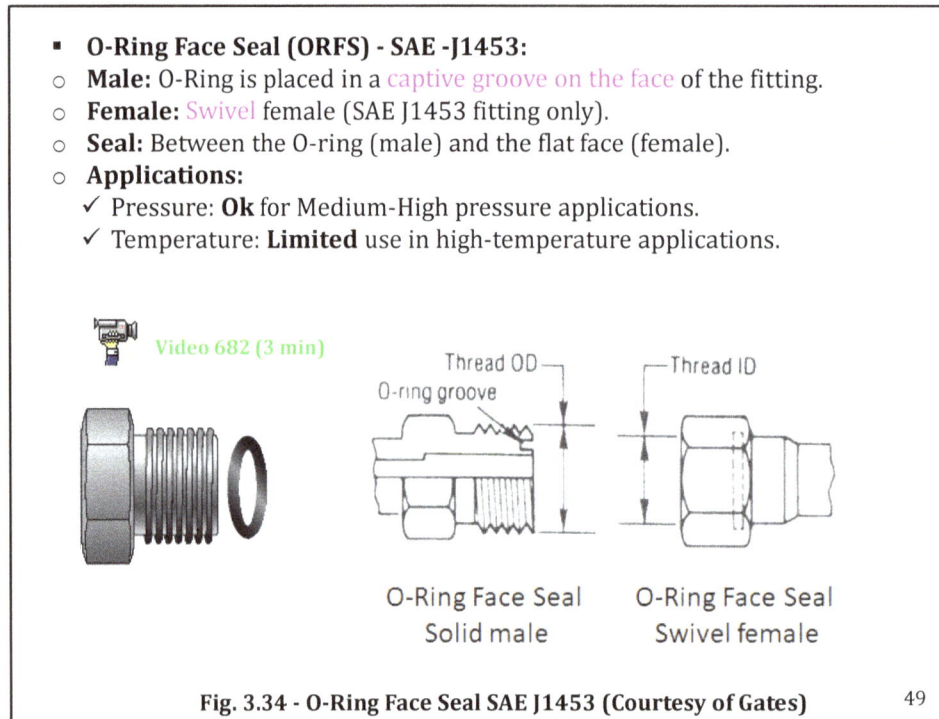

O-ring groove
Thread OD
Thread ID

O-Ring Face Seal
Solid male

O-Ring Face Seal
Swivel female

Fig. 3.34 - O-Ring Face Seal SAE J1453 (Courtesy of Gates)   49

49

**Figure 3.35 - Typical O-Ring Face Seal Fittings of Different Configurations
(Courtesy of Parker)**

❖ **British Standards:**
- Same structure as the SAE (ORB and ORFS) but the
- Thread characteristic is different.

❖ **Metric Standard (ISO/DIS 8434-3) :**
- Same structure as SAE (ORB and ORFS) but the
- Thread characteristic is different.

50

50

❖ **Example of O-Ring Face Seal (O-Lok):**
Description:
- (ORFS) fittings developed to eliminate leakage in high P systems.

Applications:
- Designed to meet the needs of mobile equipment, mining, agriculture and other heavy equipment.

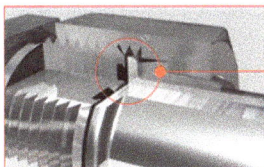

1- Tightened by Hand

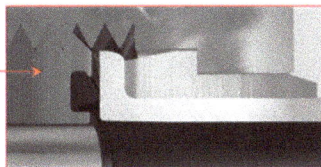

2- Tightened by Wrench to Specified Torque

3- Fluid Pressure Applies

**Fig. 3.36 - O-Lok Fitting
(Courtesy of Parker)**

51

51

Features:

- **Standard:** Meets international standards SAE J1453 and ISO8434-3.
- **P:** Up to 630 bar.
- **Material:** Steel and stainless steel.
- **Sizes:** From 1/4" to 2".
- **Lines:** Connects both tubes and hoses.
- **Assembling:** Easy to assemble.
- **Application:** Best choice if you work with construction machinery.
- **Environment:** Corrosion resistant.

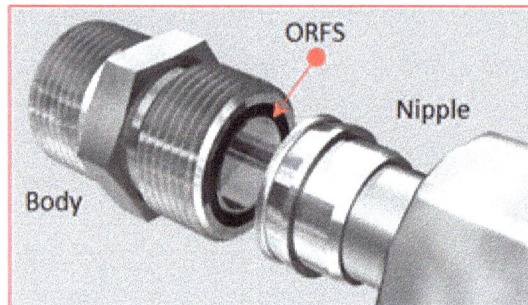

Video 653 (1.5 min)

Fig. 3.32- Contin.

52

52

3.5.5.6- Flareless Fittings with Cutting Rings

- **Designations:** "Flareless Pipe Assembly" OR "Standpipe Assembly"
- **Standards:** North American (NASP) or Metric (MSP).
- **Male:** straight threads and a 24° seat.
- **Female:** straight threads and a cutting ring for a sealing surface.
- **Seal:**
  - Male Side: Between the cutting ring and the 24° seat.
  - Female Side: Between the cutting ring and the tube.

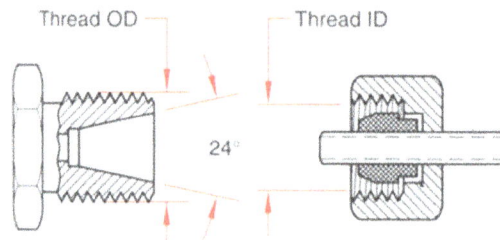

53

53

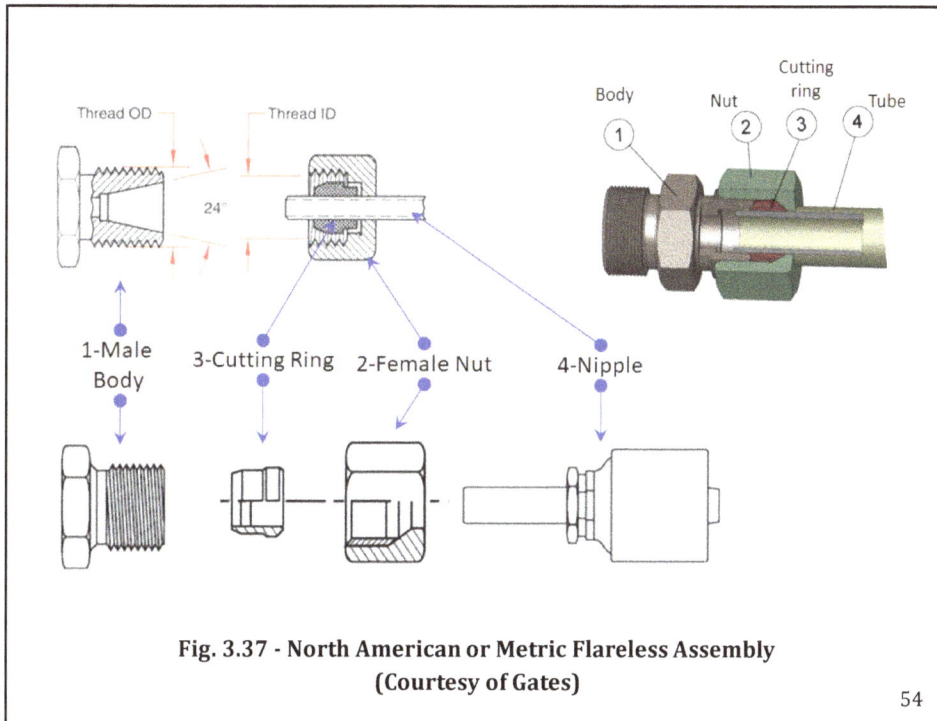

**Fig. 3.37 - North American or Metric Flareless Assembly
(Courtesy of Gates)**

54

54

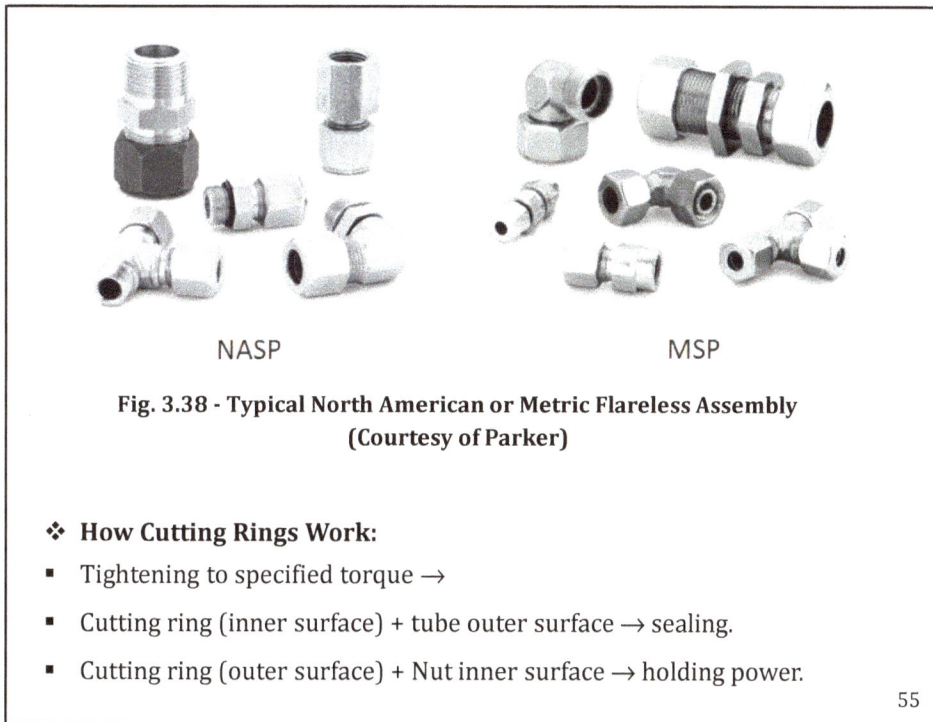

NASP   MSP

**Fig. 3.38 - Typical North American or Metric Flareless Assembly
(Courtesy of Parker)**

❖ **How Cutting Rings Work:**

▪ Tightening to specified torque →

▪ Cutting ring (inner surface) + tube outer surface → sealing.

▪ Cutting ring (outer surface) + Nut inner surface → holding power.

55

55

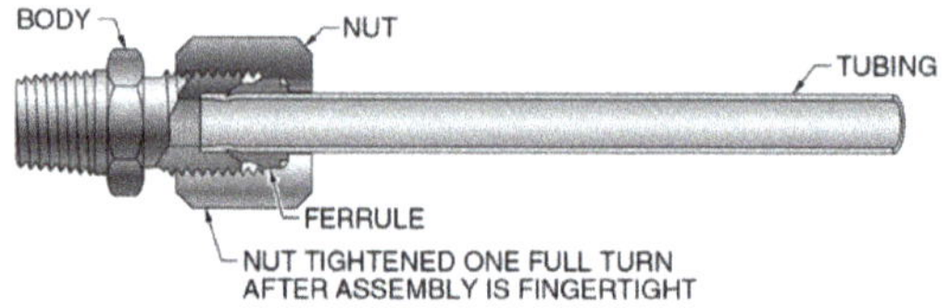

BODY — NUT — TUBING

FERRULE

NUT TIGHTENED ONE FULL TURN
AFTER ASSEMBLY IS FINGERTIGHT

Fig. 3.39 - How Cutting Rings Work (www.stauffusa.com)

Video 656 (1 min)

Video 681 (3 min)

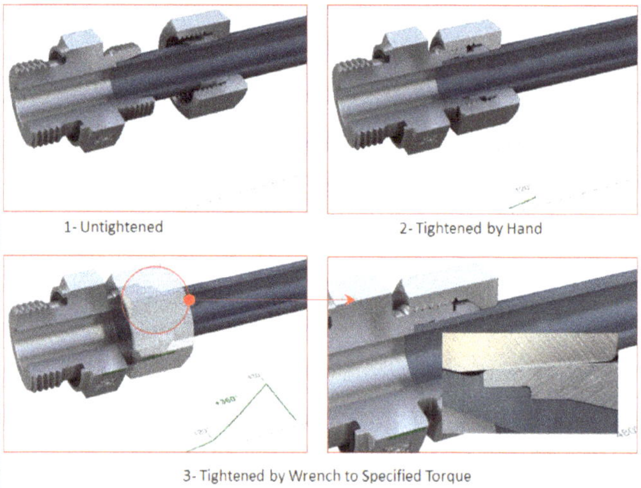

1- Untightened

2- Tightened by Hand

3- Tightened by Wrench to Specified Torque

56

3.5.5.7- Flared Fittings for Flared Tubes

Video 685 (3.5 min)

❖ **North American Standards (SAE-J514 37O):**
○ Joint Industrial Council (JIC) is defunct → Is now a part of SAE J5I4.
○ **Male/Female:** 37^0 flare seat. This male meet this female only.
○ **Seal:** is made on 37° flare seat.
○ **Thread:** straight in male and female parts.
○ **Material:** Nickel alloys, brass, carbon steel, and stainless steel.
○ **Applications:** no O-Ring → recommended for High-T applications.

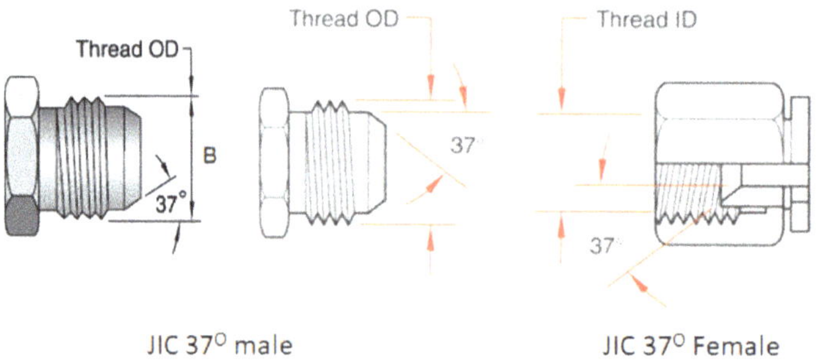

Thread OD

Thread OD

Thread ID

B

37°

37

37

JIC 37^0 male

JIC 37^0 Female

Fig. 3.40- 37^0 Flared Fitting (Courtesy of Gates)

57

- **Flare Seat Measurement:**
  - SAE 37° and SAE 45° have same sizes and same threads →
  - Carefully measure the seat angle to differentiate between the two.

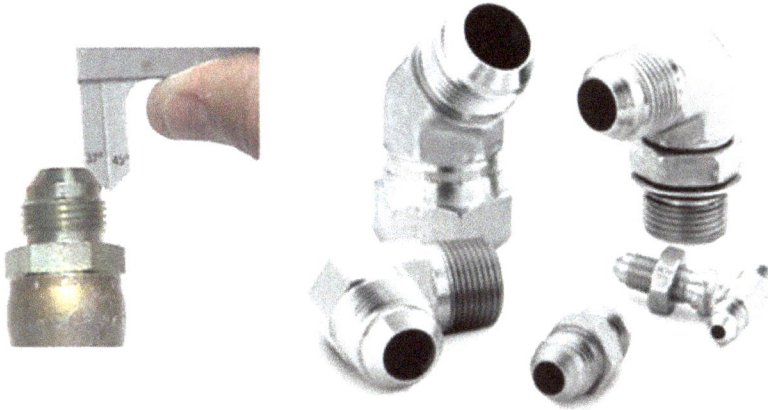

**Fig. 3.41 - Typical 37° Flared Fitting Configurations
(Courtesy of Parker)**

58

58

**Examples of Flared Fittings,
review in textbook:**

- **Figures 3.42 through 3.46**

59

59

3.5.5.8- Adaptors

- Connect fittings from foreign (British or Metric) standards to a fitting from North American National Standard:
1. Male Foreign Standard Pipe Tapered Thread to Male JIC 37° Flare.
2. Male Foreign Standard Pipe Parallel to Male JIC 37° Flare.
3. Male Foreign Standard Pipe Parallel to Male Pipe NPTF.
4. Male Foreign Standard Pipe Parallel with ORB to Male JIC 37° Flare.
5. Male Foreign Standard Pipe Parallel with ORB to Male O-Ring Flat-Face.
6. Female Foreign Standard Pipe Parallel to Male Pipe NPTF.

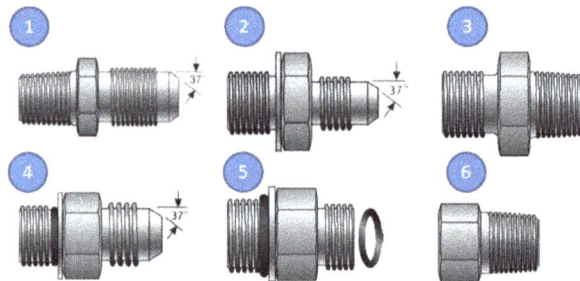

Fig. 3.47 - Adaptors to Connect Fittings from Foreign Standards to SAE Standards (Courtesy of Gates) 60

60

Fig. 3.48 - Typical Adapters (Courtesy of Parker) 61

61

3.5.5.9- Thread Characteristics Identification

❖ **By Measurements:**

Step 1: Measure the Thread Diameter:
- Use a combination O.D./I.D. Caliper.

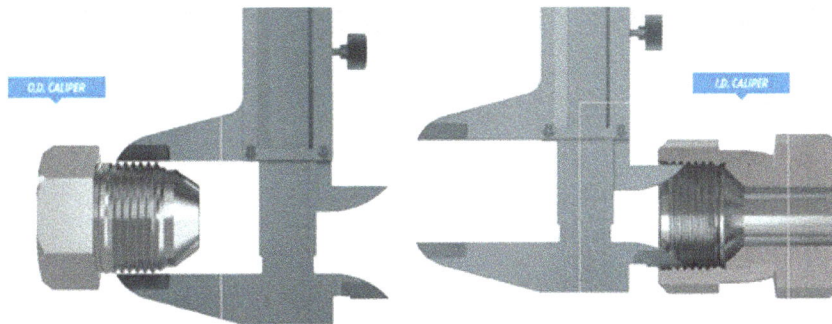

**Fig. 3.49 – Identifying Fittings by Measurements
(Courtesy of Brennan Industries)**

62

62

Step 2- Measure the Thread Pitch:
- Standard (NA) → Place the Pitch Gauge on the threads until it fits snugly.
- Standard (British and European) → count # of threads per inch.
- Standard (Metric) → measure the distance between threads.

Step 3: Measure the Seat Angle:
- Use a Seat Angle Gauge to measure the seat angle.
- Note: centerline of the fitting and the gauge must be parallel.

Note: Used fitting → worn and distorted thread → measurement error.

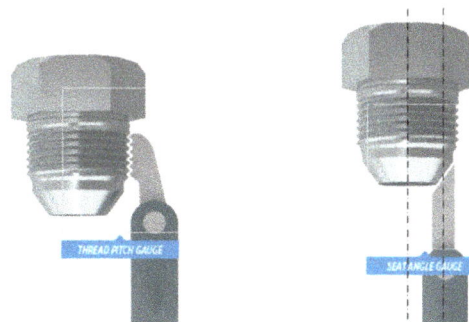

Fig. 3.49 – Continue

Video 678 (2 min)

63

63

❖ **Identifying Fittings by Thread ID Kit:**

- ▪ Faster and easier identification of fittings.
- ▪ Color coded and the size of the fitting is imprinted on the side.
- ▪ Each fitting has a female side and a male side.

Video 679 (0.5 min)   Video 680 (1 min)

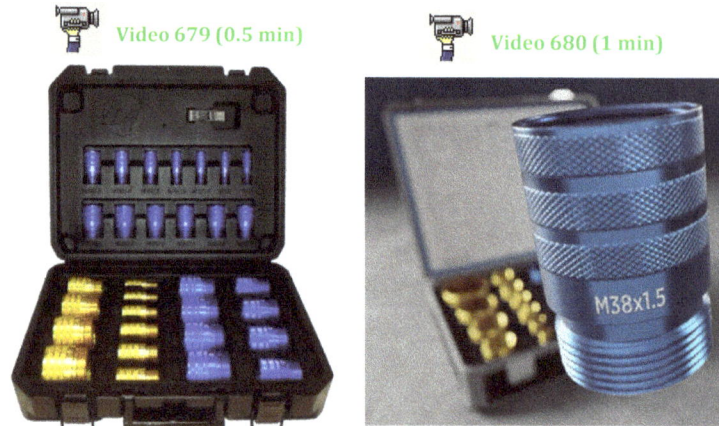

**Fig. 3.50 – Identifying Fittings by Measurements
(Courtesy of Brennan Industries)**
64

64

3.5.5.10- Swivel Joints

❖ **Conventional Swivel Joints:**
- ▪ Straight swivel joints provide rotation around one axis only.
- ▪ Elbow swivel joints provide rotation in two axes.

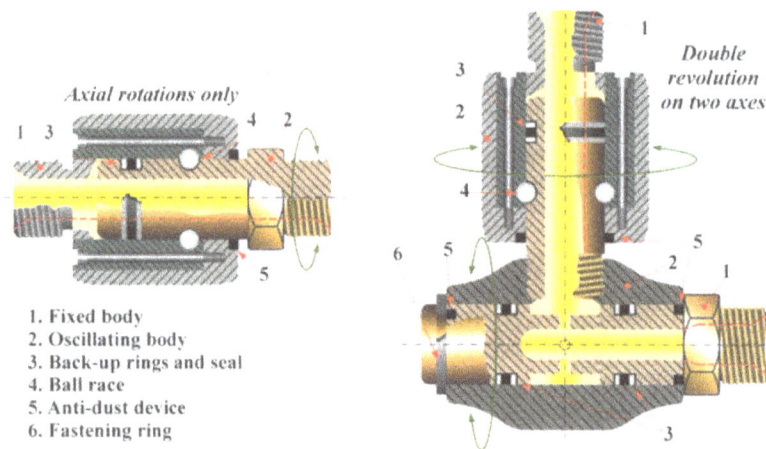

Axial rotations only

Double revolution on two axes

1. Fixed body
2. Oscillating body
3. Back-up rings and seal
4. Ball race
5. Anti-dust device
6. Fastening ring

Fig. 3.51 - Construction of Conventional Swivel Joints (Courtesy of Assufluid)
65

65

❖ **Examples** of swivel joints with the following technical characteristics:
- Variety of port options.
- Pressure capabilities up to 350 bar (5000 psi).
- Size range is 1/4" through 2".
- Full flow design minimizes pressure drop for optimum system performance.
- Nickel plating and a wide range of seal options.
- Hardened bearing races for extended service life.
- Sealed bearing design isolates bearing race from media and environment.

**Fig. 3.52 - Typical Conventional Swivel Joints
(Courtesy of Parker)**

66

66

❖ **Innovative Swivel Joints - WEO Plug-In hose fittings:**

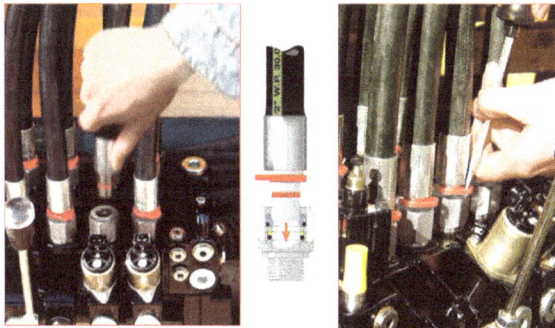

Video 430 (1 min)

Video 431 (4 min)

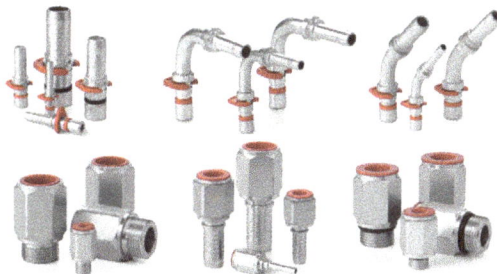

**Fig. 3.53 - WEO Plug-In
Swivel Joints
(Courtesy of CEJN)**

67

67

**More Info about WEO Plug-In Hose Fitting,
review in textbook:**

- **Figures 3.54 through 3.55**

68

3.6- Hydraulic Hoses
3.6.1- Features of Hydraulic Hoses

- Easy installation and flexible routing.
- Permit connections to components on moving parts.
- Permit misalignment between components.
- Isolate noise, shocks and vibrations.
- Require compatibility with system hydraulic fluid.
- Low equivalent bulk modulus due to the flexible walls.
- Reduce hydraulic stiffness and consequently system response.
- Usually more expensive than tubes or pipes.

69

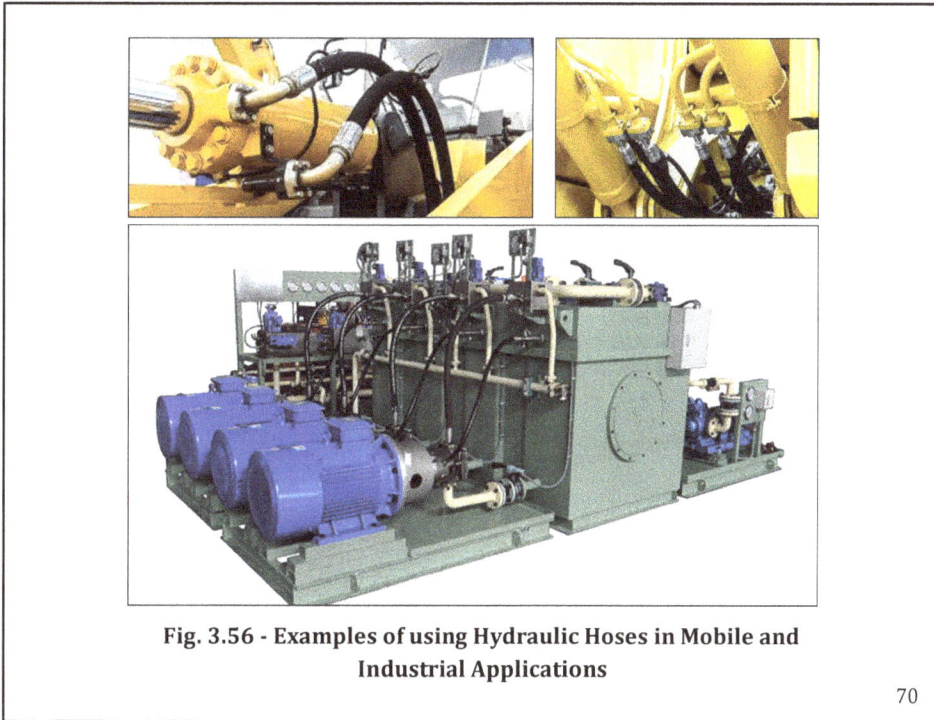

Fig. 3.56 - Examples of using Hydraulic Hoses in Mobile and Industrial Applications

70

70

3.6.2- Material and Construction of Hydraulic Hoses

Central Tube   Reinforcement Layers   Protective Cover

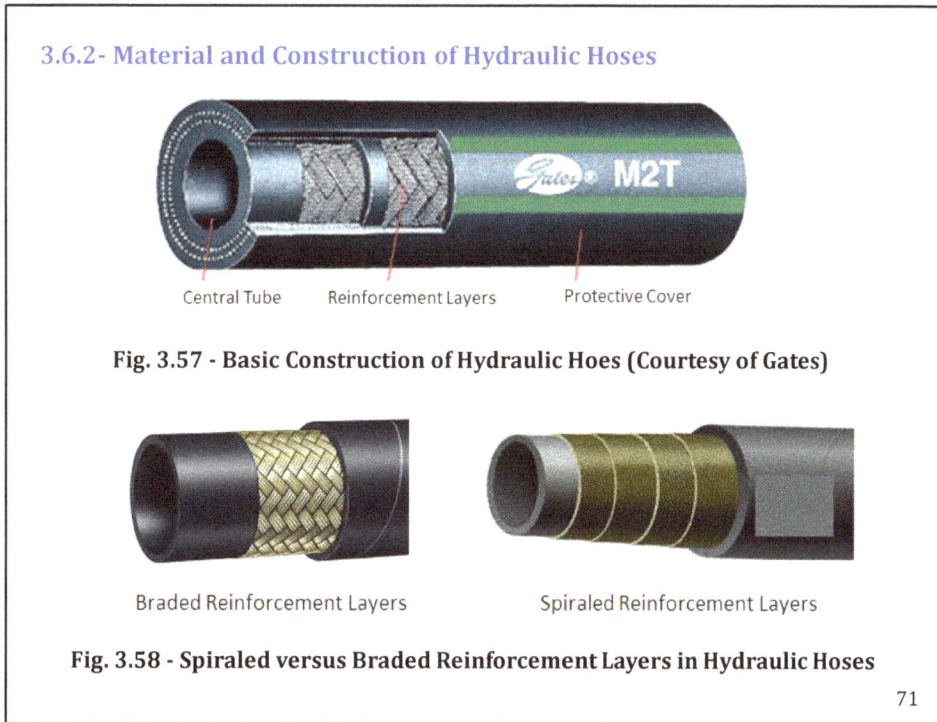

Fig. 3.57 - Basic Construction of Hydraulic Hoes (Courtesy of Gates)

Braded Reinforcement Layers   Spiraled Reinforcement Layers

Fig. 3.58 - Spiraled versus Braded Reinforcement Layers in Hydraulic Hoses

71

71

❖ **Central Tube:**
- Seamless tube.
- Fluid leakage barrier.
- Synthetic rubber or plastic.
- Compatible with the hydraulic fluid.
- Withstand the operating temperature range (low to high).

Video 161 (0.5 min)

❖ **Reinforcement Layers:**
- Braided or spiraled - textile strings or metal wires.
- Braded reinforcement layers → low to medium P.
- Spiraled reinforcement layers → medium to high P.
- Maximum allowable P. α # layers
- # layers ↑→ hose flexibility ↓.
- Braided hose is generally more flexible than spiral hose.

❖ **Outer Protective Cover:**
- Protect the hose from environmental conditions, hose abrasion, etc.
- Various materials are used such as plastic, cloth, etc.

72

72

3.6.3- Size of Hydraulic Hoses
- Unlike hydraulic pipes and tubes →
- Hose OD depends on the hose structure →
- Hose Nominal Size is based on the inner diameter "ID".
- Example is -8 means 8/16, i.e. half inch.

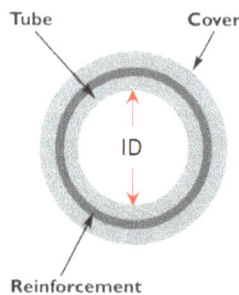

Fig. 3.59 - Nominal Size of Hydraulic Hoses (Courtesy of Gates)

Table 3.9- Size Range of Hydraulic Hoses (Courtesy of Gates)

Dash No.	Hose I.D.	
	Inches	Millimeters
3	3/16	4.8
4	1/4	6.4
5	5/16	7.9
6	3/8	9.5
8	1/2	12.7
10	5/8	15.9
12	3/4	19.0
14	7/8	22.2
16	1	25.4
20	1-1/4	31.8
24	1-1/2	38.1
32	2	50.8
40	2-1/2	63.5
48	3	76.2
56	3-1/2	88.9
64	4	101.6

73

3.6.4- Pressure Rating of Hydraulic Hoses

❖ **Standards:** ISO, SAE (most widely followed in the U.S.), DIN, etc.

❖ **SAE J517 Standard (May 1989):**

▪ **Designations:** Example (SAE100R2AT-16).

    o   **SAE100RX:** Construction and pressure rating.

    o   **A or B:** A (braided) and B (spiral) wire reinforcement layers.

    o   **T:** Indicates that no need to remove the cover to attach the fitting.

    o   **Dash number:** is the inside diameter in sixteenths of an inch.

74

74

▪ **Examples of Hoses Designations in Accordance with SAE Standard:**
- o **SAE 100R1A:** A hose with single layer of braided wire.
- o **SAE 100R2A:** A hose with double layers of braided wire.
- o **SAE 100R2B:** Two spiral plies and one braid of wire.
- o **SAE 100R4:** Wire Inserted Hydraulic Suction Hose.
- o **SAE 100R5:** Single Wire Braid, Textile Covered Hydraulic Hose.
- o **SAE 100R6:** Single Fiber Braid, (Nonmetallic), Rubber Covered Hydraulic Hose.
- o **SAE 100R7:** Thermoplastic Hydraulic Hose.
- o **SAE 100R8:** High Pressure Thermoplastic Hydraulic Hose.
- o **SAE 100R9:** High Pressure, 4-Spiral Steel Wire, Rubber Covered Hydraulic Hose.
- o **SAE 100R10:** Heavy Duty, 4-Spiral Steel Wire, Rubber Covered Hydraulic Hose.
- o **SAE 100R11:** Heavy Duty, 6-Spiral Steel Wire Rubber Covered Hydraulic Hose.
- o **SAE lOOR12:** Heavy Duty, High Impulse, 4-Spiral Wire, Rubber Cover Hydraulic Hose.
- o **SAE 100R13:** Heavy Duty, High Impulse, Multiple Spiral Wire, Rubber Covered Hose.
- o **SAE l00R14:** Hose covers for use with petroleum, synthetic, and water-base fluids.

75

75

- **Hoses Specifications in Accordance with SAE Standard:**   Video 275 (1 min)
  - A safety factor of $4 \to$ $P_{allowable} = P_{burst}/4.$
  - Maximum operating pressure↑ → OD and minimum bend radius ↑.

[¹Minimum burst pressure is 4 times maximum operating pressure.]

I.D. INCHES	DASH NO. REF.	SAE NO. & TYPE SPEC.	O.D. MAX. INCHES	MIN. BEND (INTERNAL) RAD. IN. AT MAX. OPERATING PRESSURE	MAX. OPERATING PRESSURE PSIG	I.D. INCHES	DASH NO. REF.	SAE NO. & TYPE SPEC.	O.D. MAX. INCHES	MIN. BEND (INTERNAL) RAD. IN. AT MAX. OPERATING PRESSURE	MAX. OPERATING PRESSURE PSIG
1/8	-3	100R14	0.268	1.5	1,800	3/4	-12	100R3	1.281	8	750
3/16	-3	100R1-A	0.531	3.5	3,000	3/4	-12	100R4	1.375	5	300
3/16	-3	100R1-AT	0.494	3.5	3,000	3/4	-12	100R6	1.000	6	300
3/16	-3	100R2-A&B	0.656	3.5	5,000	3/4	-12	100R7	1.125	9.5	1,250
3/16	-3	100R2-AT&BT	0.557	3.5	5,000	3/4	-12	100R8	1.300	9.5	2,250
3/16	-3	100R3	0.531	3	1,500	3/4	-12	100R9-A	1.266	9.5	3,000
3/16	-3	100R5	0.539	3	3,000	3/4	-12	100R9-AT	1.255	9.5	3,000
3/16	-3	100R6	0.489	2	500	3/4	-12	100R10-A	1.469	11	5,000
3/16	-3	100R7	0.450	3.5	3,000	3/4	-12	100R10-AT	1.450	11	5,000
3/16	-3	100R8	0.575	3.5	5,000	3/4	-12	100R11	1.594	11	6,250
3/16	-3	100R10-A	0.781	4	10,000	3/4	-12	100R12	1.241	9.5	4,000
3/16	-3	100R11	0.906	4	12,500	3/4	-12	100R13	1.306	9.5	5,000
3/16	-4	100R14	0.324	2	1,800	3/4	-14	100R14	0.917	9	800
1/4	-4	100R1-A	0.656	4	2,750	7/8	-14	100R1-A	1.344	11	1,125
1/4	-4	100R1-AT	0.557	4	2,750	7/8	-14	100R1-AT	1.252	11	1,125
1/4	-4	100R2-A&B	0.719	4	5,000	7/8	-14	100R2-A&B	1.406	11	2,000
1/4	-4	100R2-AT&BT	0.619	4	5,000	7/8	-14	100R2-AT&AB	1.315	11	2,000
1/4	-4	100R3	0.594	3	1,250	7/8	-16	100R5	1.266	7.38	800
1/4	-4	100R5	0.601	3.38	3,000	7/8	-16	100R14	1.081	9	800
1/4	-4	100R6	0.531	2.5	400	1	-16	100R1-A	1.547	12	1,000
1/4	-4	100R7	0.538	4	2,750	1	-16	100R1-AT	1.440	12	1,000
1/4	-4	100R8	0.660	4	5,000	1	-16	100R2-A&B	1.609	12	2,000
1/4	-4	100R10-A	0.844	4	8,750	1	-16	100R2-AT&BT	1.531	12	2,000
1/4	-4	100R11	0.969	5	11,250	1	-16	100R3	1.547	8	565
1/4	-5	100R14	0.397	3	1,500	1	-16	100R4	1.625	6	250
5/16	-5	100R1-A	0.719	4.5	2,500	1	-16	100R7	1.445	12	1,000
5/16	-5	100R1-AT	0.619	4.5	2,500	1	-16	100R8	1.520	12	2,000
5/16	-5	100R2-A&B	0.781	4.5	4,250	1	-16	100R9-A	1.609	12	3,000
5/16	-5	100R2-AT&BT	0.682	4.5	4,250	1	-16	100R9-AT	1.594	12	3,000
5/16	-5	100R3	0.719	4	1,200	1	-16	100R10-A	1.797	14	4,000
5/16	-5	100R5	0.695	4	2,250	1	-16	100R10-AT	1.790	14	4,000
5/16	-5	100R6	0.594	3	400	1	-16	100R11	1.953	14	5,000
5/16	-5	100R7	0.616	4.5	2,500	1	-16	100R12	1.542	12	4,000

Table 3.10 – Hoses Specifications in accordance with SAE Standard   76

76

- **Fluid Compatibility of Hydraulic Hoses in Accordance with SAE Standard:**

SAE DIMENSIONAL AND PERFORMANCE STANDARDS FOR HYDRAULIC HOSE							
SAE standard hydraulic hose type/application	Compatible hydraulic fluids	Temperature range, °F	ID range, in.	Maximum operating pressure, psi	Proof pressure range, psi	Minimum burst pressure range, psi	Minimum bend radius, in.
100R1—Steel wire reinforced, rubber coated	Petroleum & water based	-40 to 212	3/16 to 2	575 to 3,250	1,150 to 6,500	2,300 to 13,000	3.5 to 25
100R2—High-pressure steel wire, reinforced rubber cover	Petroleum & water based	-40 to 212	3/16 to 2	1,150 to 6,000	2250 to 12,000	4,500 to 24,000	3.5 to 25
100R3—Double fiber, braid rubber cover - High-temp, low-pressure	Petroleum & water based	-40 to 212	3/16 to 1-1	375 to 1,500	750 to 3,000	1,500 to 6,000	3 to 10
100R4—Wire inserted, hydraulic suction and return	Petroleum & water based	-40 to 212	3 to 4	35 to 300	70 to 600	140 to 1,200	5 to 24
100R5—Single wire braid, textile cover; Transportation/DOT	Petroleum & water based	-40 to 212	3/16 to 3-1/16	200 to 3,000	400 to 6,000	800 to 12,000	3 to 33
100R6—Single fiber braid, rubber cover—Transportation	Petroleum & water based	-40 to 212	3/16 to 3	300 to 500	600 to 1,000	1,200 to 2,000	2 to 6
100R7—Single fiber braid, thermoplastic-Hydraulic	Petroleum, water based & synthetic	-40 to +212	1/8 to 1	1,000 to 3,000	2,000 to 6,000	4,000 to 12,000	1 to 12
100R8—High pressure, thermoplastic-Hydraulic	Petroleum, water based & synthetic	-40 to 212	1/8 to 1	2,000 to 6,000	4,000 to 12,000	8,000 to 24,000	1 to 12
100R9	No longer part of the SAE standard.						
100R10	No longer part of the SAE standard.						
100R11	No longer part of the SAE standard.						

Table 3.11 - SAE Standard for Hydraulic Hoses
(Hydraulic & Pneumatic Magazine)   77

77

❖ **ISO 18752 Standard** (released in 2006):
- Large OEMs switched to ISO standards
- → Ensure global sale and service of their equipment.
- MWP = Max Workable Pressure

Class	35	70	140	210	250	280	350	420	560
MWP[a] (bar)	35	70	140	210	250	280	350	420	560
MWP[a] (MPa)	3.5	7	14	21	25	28	35	42	56
MWP[a] (psi)	500	1000	2000	3000	3500	4000	5000	6000	8000

Nominal Size ISO / Inch	35	70	140	210	250	280	350	420	560
5 / -3	•	•	•	•	•	•	•	•	N/A
6.3 / -4	•	•	•	•	•	•	•	•	N/A
8 / -5	•	•	•	•	•	•	•	•	N/A
10 / -6	•	•	•	•	•	•	•	•	N/A
12.5 / -8	•	•	•	•	•	•	•	•	N/A
16 / -10	•	•	•	•	•	•	•	•	•
19 / -12	•	•	•	•	•	•	•	•	•
25 / -16	•	•	•	•	•	•	•	•	•
31.5 / -20	•	•	•	•	•	•	•	•	•
38 / -24	•	•	•	•	•	•	•	•	N/A
51 / -32	•	•	•	•	•	•	•	•	N/A
63 / -40	•	•	N/A	N/A	N/A	N/A	N/A	N/A	N/A
76 / -48	•	N/A	N/A	N/A	N/A	N/A	N/A	N/A	N/A
102 / -64	•	N/A	N/A	N/A	N/A	N/A	N/A	N/A	N/A

Note: ● = Applicable   N/A = Not Applicable   ᵃ = Maximum Working Pressure

Table 3.12 - ISO 18752 Standard Pressure Classes for Hydraulic Hoses (Courtesy of Parker)

78

78

❖ **Resistance to Impulse:**
- 4 grades (A, B, C, and D) based on (# cycles + P pulse + T).
- Each grade is designated as Standard (S) or Compact (C).
- Compact types have a smaller OD and smaller bend radius.

Grades and Types				
		Resistance to Impulse		
Grade	Type[a]	Temperature °C	Impulse Pressure (% of MWP[b])	Minimum Number of Cycles
A	AS	100	133%	200,000
	AC			
B	BS	100	133%	500,000
	BC			
C	CS	120	133% and 120%[c]	500,000
	CC			
D	DC	120	133%	1,000,000

[a] Standard or compact, e.g. CS is grade C and standard type.
Standard types have larger outside diameters and larger bend radii and compact types have smaller outside diameters and smaller bend radii.
[b] Maximum working pressure.
[c] 120% of the MWP shall be used for classes 350, 420 and 560 instead of 133%.

Table 3.13 - ISO 18752 Standard Pressure Classes for Hydraulic Hoses (Courtesy of Parker)

79

79

3.6.5- Hydraulic Hose Bend Radius

- Hose ID $\uparrow\rightarrow$ Bend Radius$\uparrow$
- Installation exceeds minimum bend radius
- $\rightarrow$ kinks, hose life $\downarrow$, premature failure.

387
Hydraulic – Constant Working Pressure
ISO 18752

# Part Number 387	Standard Cover 387	Tough Cover 387TC	Super Tough 387ST	Hose I.D.		Hose O.D.		Working Pressure		Minimum Bend Radius		Weight		Vacuum Rating		Parkrimp 43 Series	Parkrimp 77 Series
	ISO 18752 Performance			inch	mm	inch	mm	psi	MPa	inch	mm	bs/ft	kg/m	inches of Hg	kPa		
387-4	AC	AC	AC	1/4	6,3	0.53	13,4	3000	21,0	2	50	0.16	0,24	24	80	•	
387-6	AC	AC	AC	3/8	10	0.69	17,4	3000	21,0	2-1/2	65	0.23	0,34	24	80	•	
387-8	AC	AC	AC	1/2	12,5	0.82	20,7	3000	21,0	3-1/2	90	0.29	0,43	24	80	•	
387-10	AC	AC	AC	5/8	16	0.94	23,9	3000	21,0	4	100	0.33	0,49	24	80	•	
387-12	AC	AC	AC	3/4	19	1.10	27,8	3000	21,0	4-3/4	120	0.58	0,86	24	80	•	
387-16	AC	AC	AC	1	25	1.40	35,4	3000	21,0	6	150	0.79	1,17	24	80	•	
387-20	BC	CC	CC	1-1/4	31,5	1.82	46,3	3000	21,0	8-1/4	210	1.74	2,59	18	60	•	•
387-24	BC	CC	CC	1-1/2	38	2.08	52,8	3000	21,0	10	250	2.01	2,99	18	60		•
387-32	BC	CC	CC	2	51	2.61	66,2	3000	21,0	12-1/2	320	2.75	4,09	18	60		•

Table 3.14 - Example of Minimum Bend Radius Reported by the Hose Manufacturer (Courtesy of Parker)

80

- **Minimum Straight Length:**
- When bending a hose assembly $\rightarrow$
- hose bend starts at a minimum straight length = 2 OD.

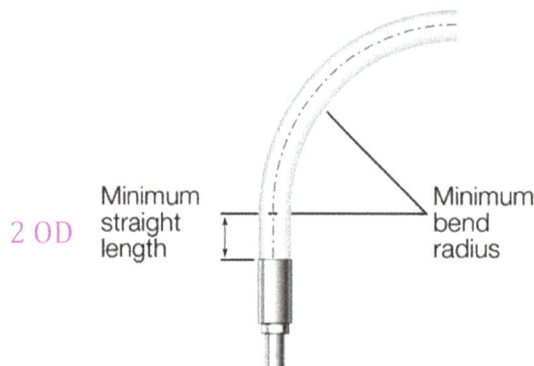

Fig. 3.60 - Minimum Straight Length for Bent Hoses

81

3.6.6- Hydraulic Hose Assembly

- C1 and C2 are the lengths of the two hose couplings.

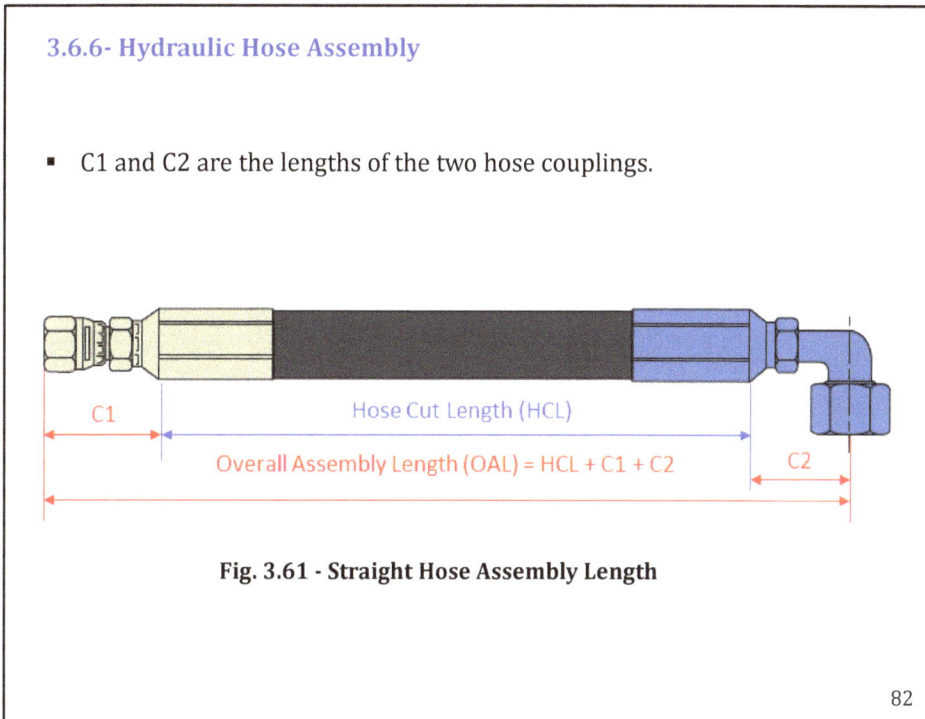

Fig. 3.61 - Straight Hose Assembly Length

82

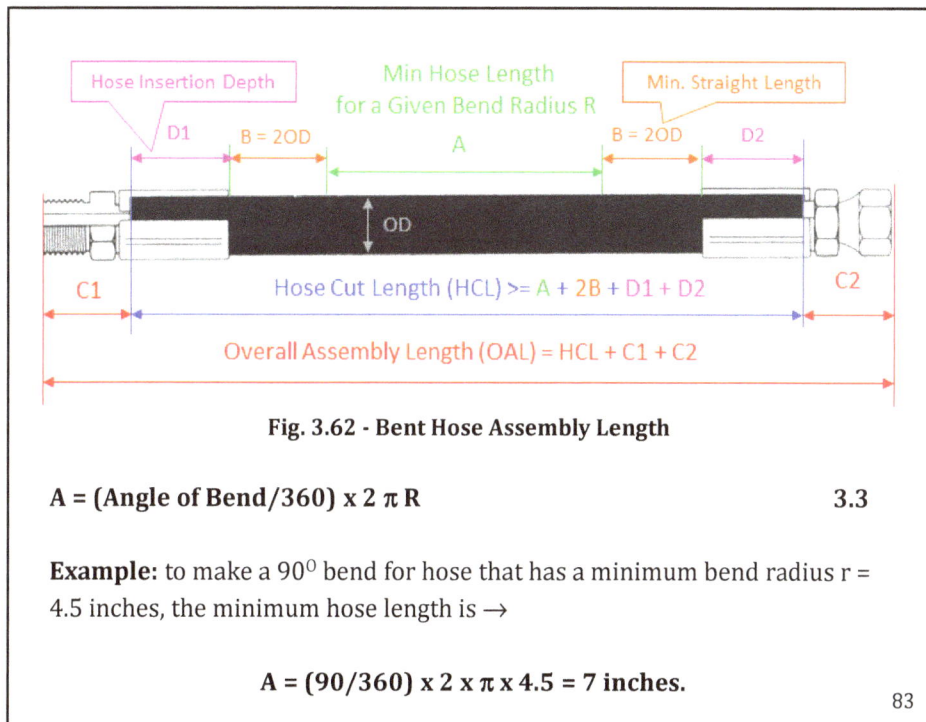

Fig. 3.62 - Bent Hose Assembly Length

A = (Angle of Bend/360) x 2 π R   3.3

Example: to make a 90⁰ bend for hose that has a minimum bend radius r = 4.5 inches, the minimum hose length is →

A = (90/360) x 2 x π x 4.5 = 7 inches.

83

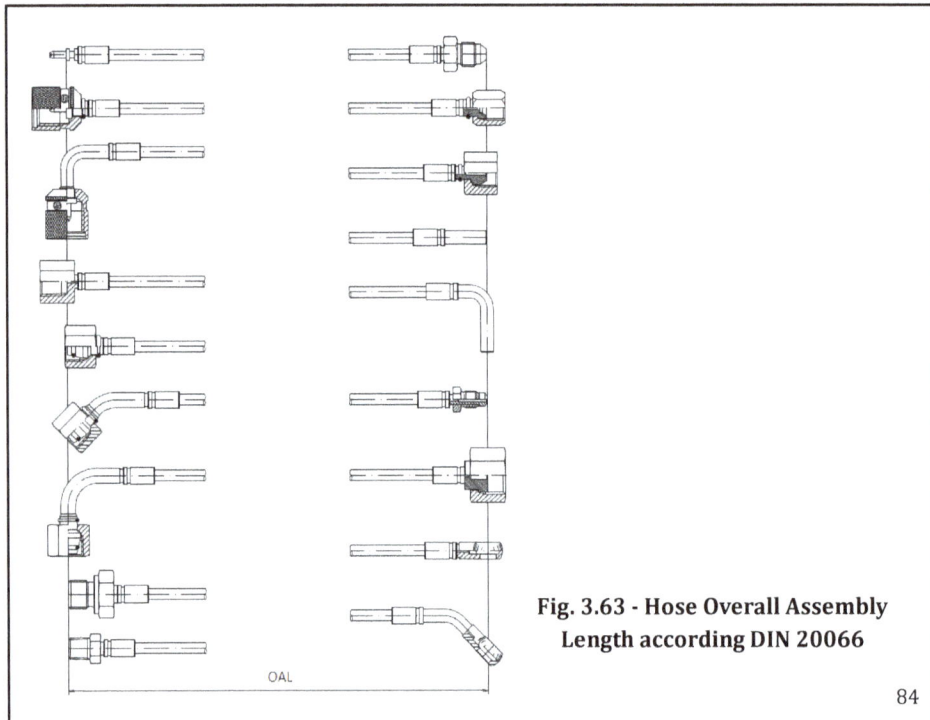

Fig. 3.63 - Hose Overall Assembly Length according DIN 20066

84

84

3.6.7- Hose Couplings

3.6.7.1- Identification of Hose Couplings

- Two end joints are called Hose Couplings.
- **Hose End:** for hose attachment.
- **Thread End** (or adaptor): for port.

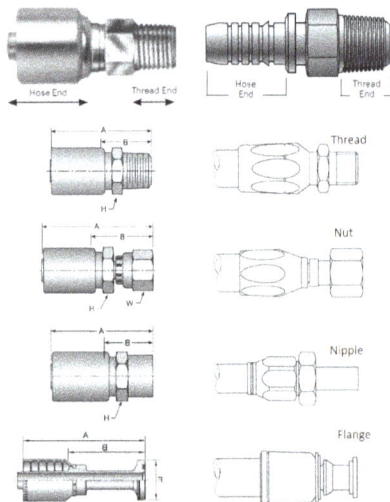

Fig. 3.64 - Identification of Hose Couplings (Courtesy of Gates)

85

85

3.6.7.2 - Reusable versus Permanent Couplings

❖ **Reusable Couplings:**

- Emergency repairs.
- Assembled by hand wrenches; there is no crimper required.
- Aren't recommended if there are high pulsating pressure.

Fig. 3.65

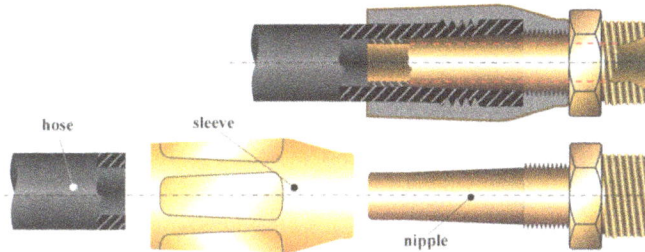

Fig. 3.66 - Reusable Hose Couplings (Courtesy of Assufluid)

86

86

❖ **Permanent Couplings:**

- Requires crimping.
- Nipple ensures seizing and prevent hose narrowing.

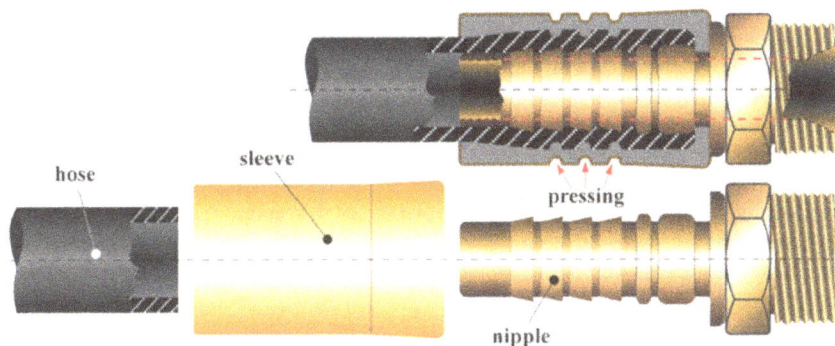

Fig. 3.67 - Permanent Hose Couplings (Courtesy of Assufluid)

87

87

- Permanent hose couplings:

Preassembled   Two-piece configurations

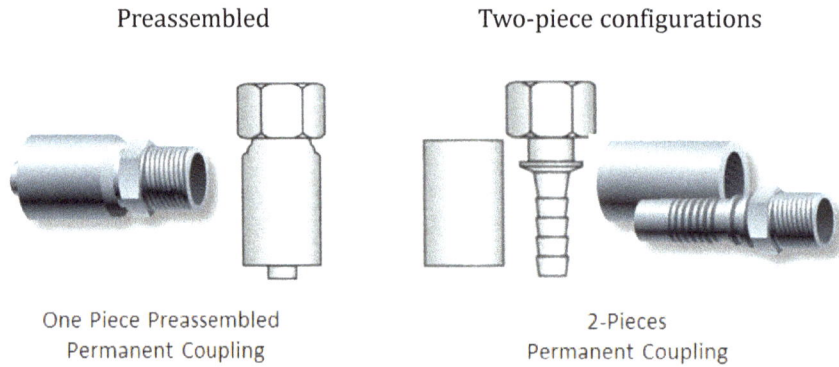

One Piece Preassembled   2-Pieces
Permanent Coupling   Permanent Coupling

Fig. 3.68 - Permanent Hose Couplings (Courtesy of Gates)

88

88

3.6.7.3- Hose Couplings Configurations

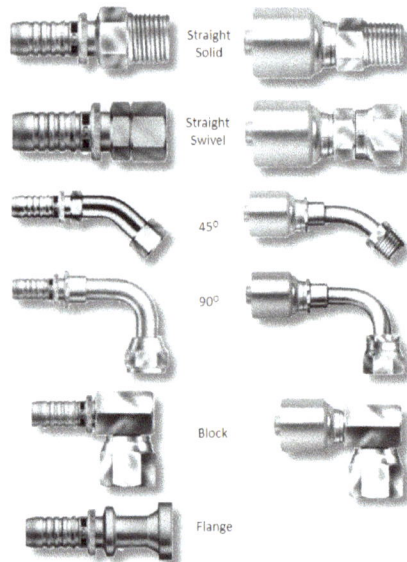

Straight Solid

Straight Swivel

45°

90°

Block

Flange

Fig. 3.69 - Hose Coupling Common Configurations (Courtesy of Gates)

89

89

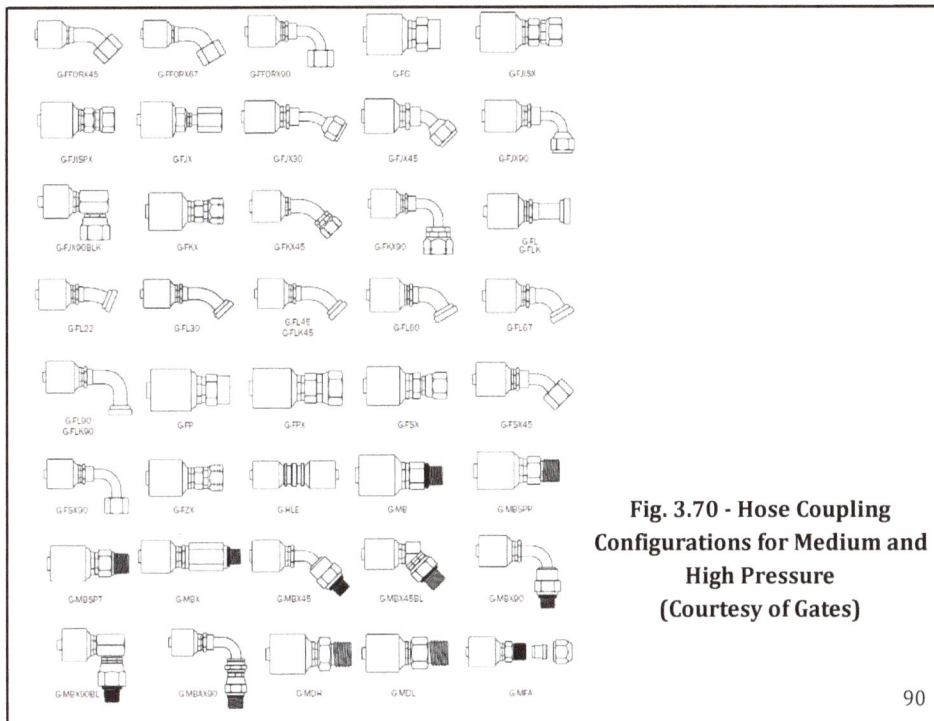

Fig. 3.70 - Hose Coupling Configurations for Medium and High Pressure (Courtesy of Gates)

90

90

3.6.7.4- Hose Crimping

❖ **Crimpable Coupling:**

▪ One-piece crimpable hose coupling.

▪ Teeth in the crimpable fittings bite down to the hose wire

▪ → metal-to-metal grip → maximum integrity.

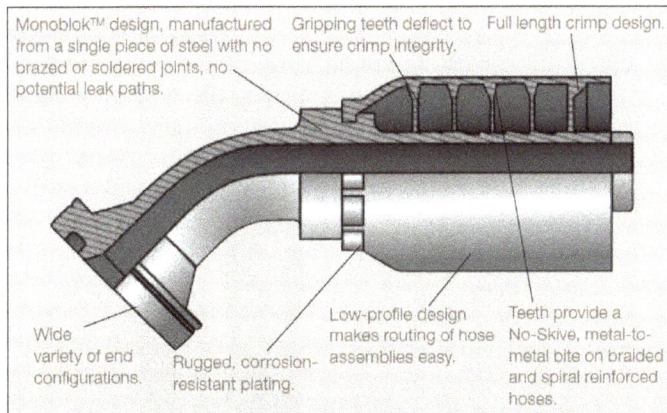

Fig. 3.71 - Typical Design of Preassembled Cimpable Hose Coupling (Courtesy of Parker)

91

91

❖ **Crimping Machines:**
- A crimper is a hydraulic ram.
- Fluid pressure → extend the ram → crimp the fittings.
- They are available in various styles and power capacity.

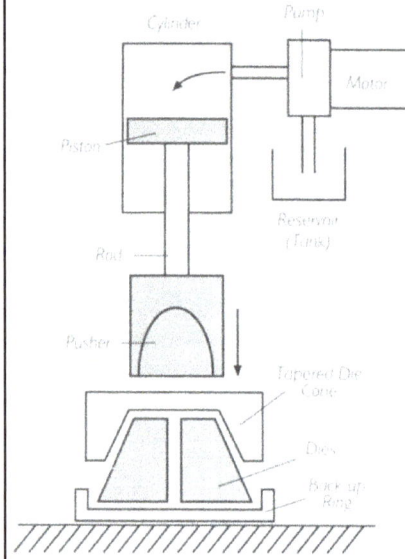

**Fig. 3.72 - Hose Crimping Machine
(Courtesy of Gates)**

92

92

❖ **Hose Crimping Method:**
- **Notes:**

o Never re-crimp couplings.

o Some hose design requires removing the upper part of the rubber cover at the hose end so that the fitting can be installed.

o A specific crimping machine → instructions provided.

Fig. 3.73 - Prepare a Hose for Crimping

93

93

❖ **Measuring Crimp Diameters:**
▪ Crimp Diameter is measured:
   o Between the ridges (caliper fingers do not touch the ridges).
   o Halfway between the top and bottom of the coupling.

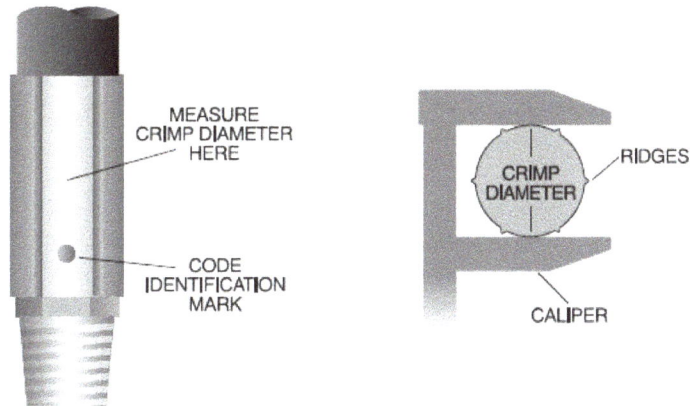

Fig. 3.74 – Measuring Crimp Diameter (Courtesy of Gates)

94

94

▪ **General Crimping Procedure:**

A specific crimping machine → instructions provided.

More Info about Crimping Process review in textbook:

▪ **Table 3.10**

95

95

3.6.8- Quick Connect Couplings

Video 749

9 Different Types A-H

A- Conventional Quick Connect Coupling:

- **Definition:** A Quick Connect Coupling → quickly connect and disconnect hoses without the use of tools or special devices.
- **Standard:** ISO 16028.
- **Flat Face Quick Connect** coupling consists of:
  - **Coupler:** Other nomenclature are "Female", "Socket", "Body".
  - **Nipple:** Other nomenclature are "Male", "Plug", "Adapter".

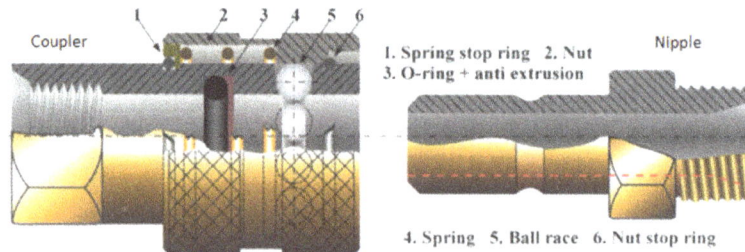

Coupler

1. Spring stop ring 2. Nut
3. O-ring + anti extrusion

Nipple

4. Spring   5. Ball race   6. Nut stop ring

Fig. 3.75- Flat Face Quick Connect Coupling (Courtesy of Assufluid)   96

96

- **Connection:**
  - Push the movable part (usually the coupler)
  - Towards the fixed part (usually the nipple).
- **Disconnection:** Is performed by pulling the nut in the coupler backward.

- **Residual Pressure:** Such couplings are used for lines where no residual pressure or fluid leakage is found upon connection or disconnection.
- **Pressure Drop:** (5-10 bar) or more if not sized properly.
- **Sizing:** These couplings should be sized based on the internal flow area of the coupling.

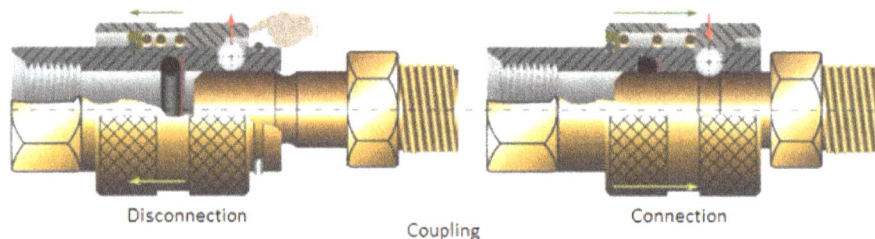

Disconnection

Coupling

Connection

Fig. 3.75- Continue

97

97

B- **Quick Connect Coupling with Nonreturn Valve:**

- A single non-return valve (7) in the female side →
- During disconnection → perfect sealing on female side if pressurized.

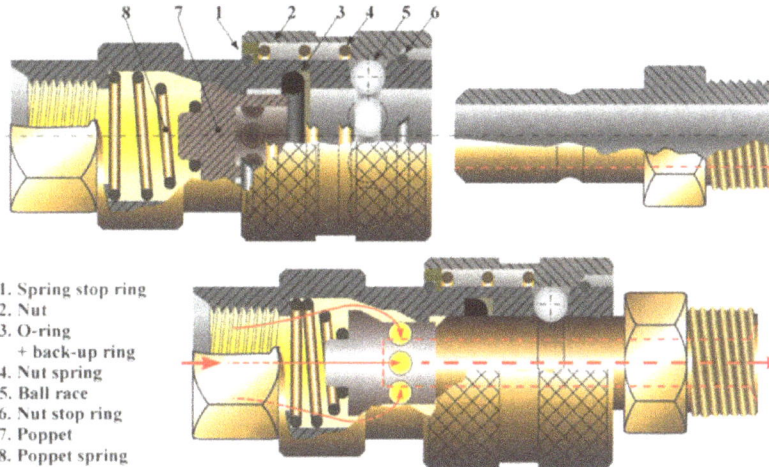

1. Spring stop ring
2. Nut
3. O-ring
   + back-up ring
4. Nut spring
5. Ball race
6. Nut stop ring
7. Poppet
8. Poppet spring

**Fig. 3.76 - Quick Connect Coupling with Single Non-Return Valve
(Courtesy of Assufluid)**   98

98

- Two non-return valves, (7) in the female side & (9) in the male side →
- During disconnection → perfect sealing on both male and female sides.
- Such a quick couplings work under pressure 400-600 bar (6000-9000) psi.

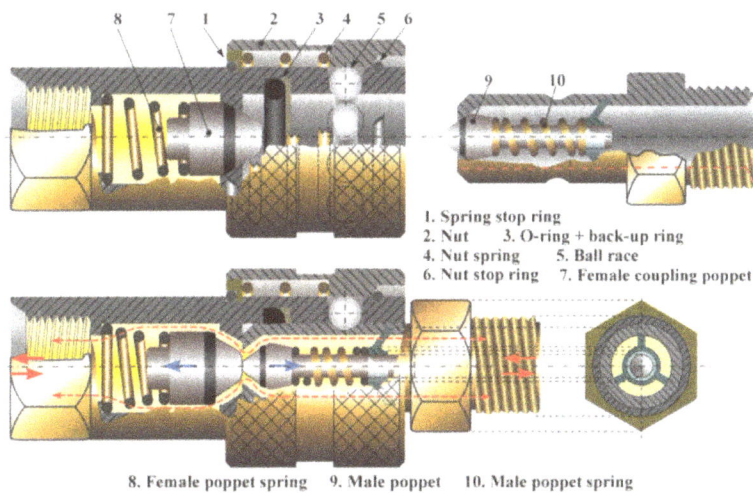

1. Spring stop ring
2. Nut   3. O-ring + back-up ring
4. Nut spring   5. Ball race
6. Nut stop ring   7. Female coupling poppet

8. Female poppet spring   9. Male poppet   10. Male poppet spring

**Fig. 3.77 - Quick Connect Coupling with Double Non-Return Valves
(Courtesy of Assufluid)**   99

99

C- Innovative iLok™ Coupling: Video 285 (6.5 min)

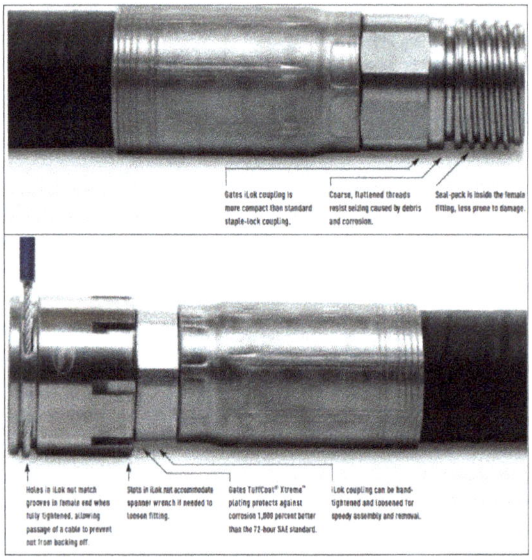

**Fig. 3.78 - Male and Female Parts of iLok™ Quick Connect Coupling
(Courtesy of Gates)**

100

100

**More Info about iLok Coupling
review in textbook:**

- **Figure 3.79**

101

101

<u>D-</u> **Innovative TLX Coupling:**

- Same concept is been developed by another manufacturer under a brand name of TLX Quick Connect Coupling.

Video 427 (0.5 min)   Video 428 (1 min)   Video 429 (1 min)

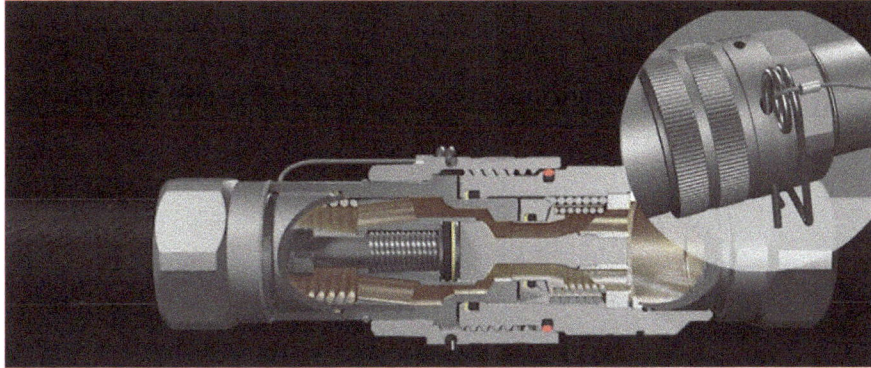

**Fig. 3.80 - Locking Mechanism of TLX Quick Connect Coupling
(Courtesy of CEJN)**

102

102

<u>E-</u> **Universal Push-to-Connect (UPTC) Assembly:**   Video 661 (1.5 min)

- **Description:**
- o Reliable leak-free connections for hose and tube assembly.
- o Utilizes standard O-Ring Face Seal (ORFS) or EO (24° DIN cone) fittings.
- o Suitable for hoses (rubber or thermoplastic) and tube (inch or metric).

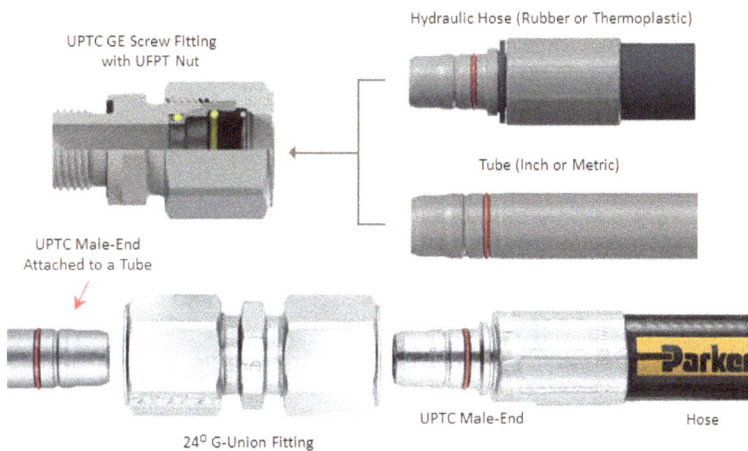

Fig. 3.81 - Universal Push-to-Connect (UPTC) (Courtesy of Parker)   103

103

**More Info about UPTC Coupling
review in textbook:**

- **Figure 3.82 through 3.84**

104

104

F- Push-Lok Assembly for Low Pressure:

Video 663 (3.5 min)

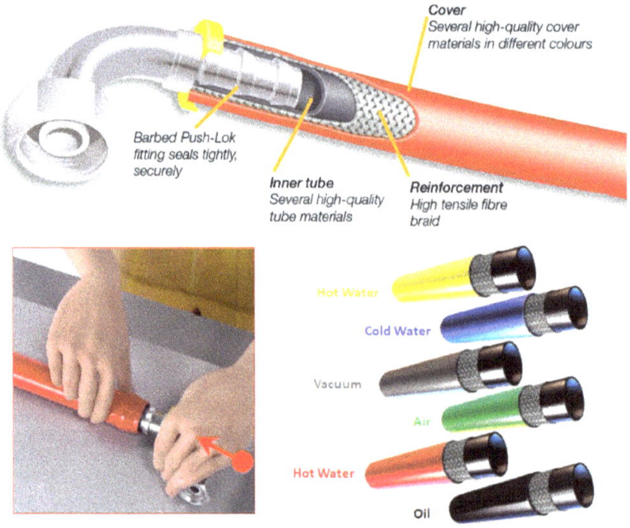

Fig. 3.85 - Push-Lok Hose and Fitting Assembly (Courtesy of Parker)

105

105

**More Info about Push-Lok Assmbly
review in textbook:**

- **Figure 3.86**

106

106

G- Innovative Quick Connect System for Low Pressure:

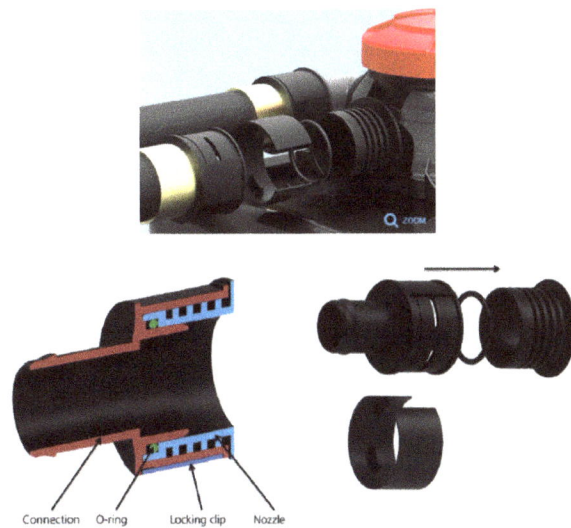

Connection   O-ring   Locking clip   Nozzle

Fig. 3.87 - Innovative Quick Connect Coupling (Courtesy of ArgoHyots)

107

107

H- CEJN Multi-X Quick Connect Couplings:

- **Features:**
- Are designed to easy connect multiple lines simultaneously.
- Great flexibility, high performance, and trouble-free operation.
- Are designed to meet and exceed requirements for challenging mobile hydraulic application.

Fig. 3.88 - Multi-X Quick Connect Coupling (Courtesy of CEJN)

108

108

Video 425 (1 min)

EASY TO MANEUVER

ENVIRONEMENT FRIENDLY FLAT-FACE CONNECTIONS

CONNECTABLE UNDER PRESSURE

TORSION FREE CONNECTORS

FLEXIBLE INSTALLATION

SMALL SPACE REQUIREMENTS

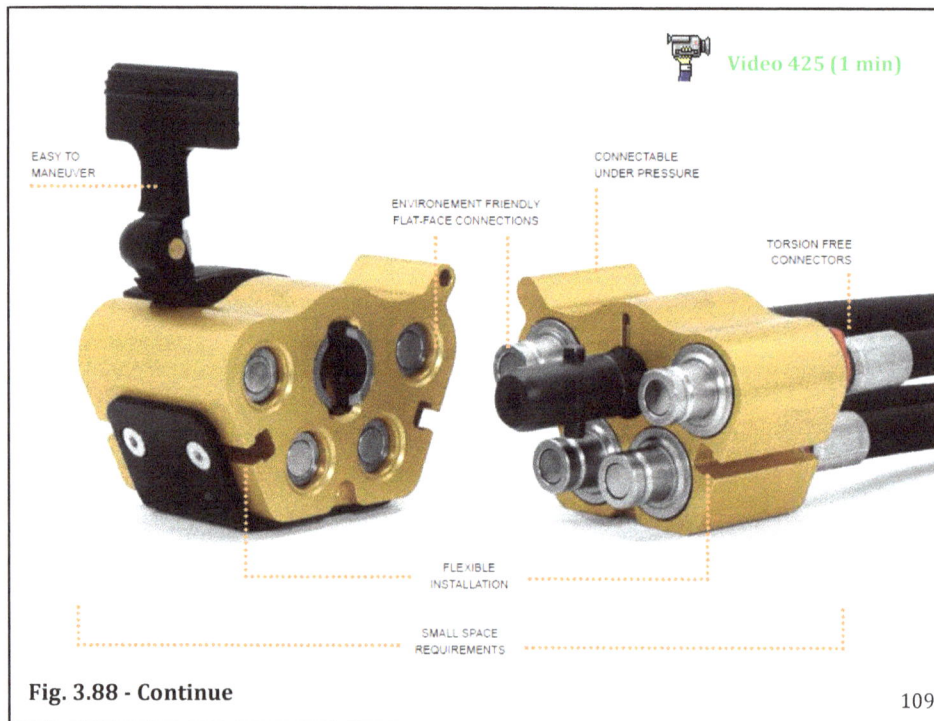

Fig. 3.88 - Continue

109

109

- **Concept of Operation:**

**Fig. 3.89 - Conceptual Operation of
Multi-X Quick Connect Coupling
(Courtesy of CEJN)**

Working pressure of up to 350 bar (5000 psi) can be used on half the port simultaneously with the other half of the port used as return lines with a maximum pressure 50 bar (725 psi).

Both the female and male plate can be used in the fixed part.

Video 426 (0.5 min)

Electric connectors can also be easily attached.

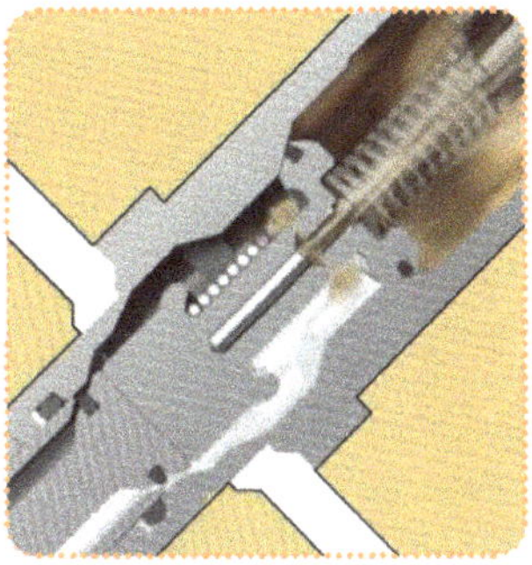

Pressure is connected without spillage making it simple to connect with residual pressure in the system

110

110

3.6.9- Hose Guards

❖ Using Hose Guards is recommended for the following reasons:

- **Hose Protection:**
  o Against harsh environmental and abrasion → prolong hose service life.

- **Operator Protection:**
  o Protect machine operators against whiplash and oil injection hazards.

- **Machine Protection:**
  o Prolonged hose life → reduced costly machine downtime and liabilities.

111

111

❖ Hose guards are available in several styles and sizes as follows:

- **Metallic Spring Guards and Armor Guards:**
  o Are constructed of steel wire.
  o Plated to resist rust.

Steel Spring Guard

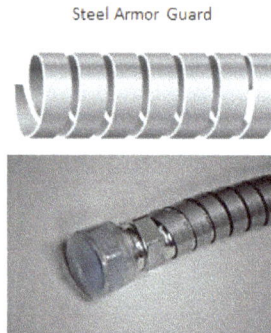

Steel Armor Guard

sapphirehydraulics.com

Fig. 3.90 - Metallic Spring and Armor Guards (Courtesy of Parker)

112

112

- **Nonmetallic Spiral Wrap Hose Guards:**
  o Are resistant to oil, lubricants, gasoline, most solvents.
  o Withstand ambient temperatures from -40° to +300° F.
  o Available in different colors →
  o High-pressure hoses (red) and Low-pressure hoses (blue).

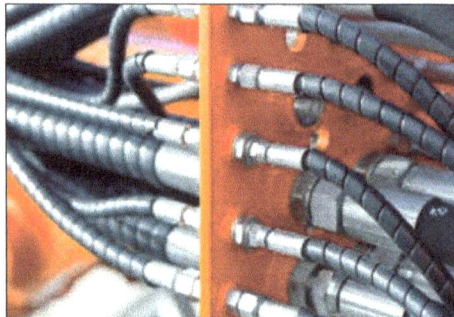

**Fig. 3.91- Nonmetallic Spiral Wrap Guards
(sapphirehydraulics.com)**

113

113

- **Hose Shields:**
  - o <u>Material:</u> abrasion and hydrocarbon resistant material.
  - o <u>Features:</u> Durable, resistant to solvents, oils, grease, and gasoline.

Fig. 3.92- Hose Shields
(www.epha.com)

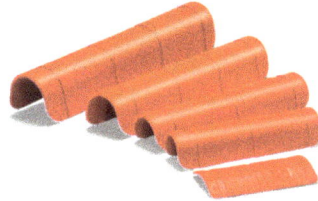

- **Hose Protection Sleeves:**
- Protect hoses from contamination, direct sunlight, and environment.
- They are made of heat and chemical resistant materials.

Fig. 3.93- Hose Protection Sleeves
(www.sealsaver.com)

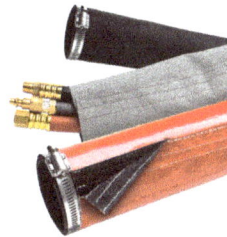

114

114

- **Hose Oil-Injection Protection Sleeves:**
  - o A pinhole leak in a hydraulic hose under pressure →
  - o Fluid injection through human body →
  - o Loss of organs or even life if not treated immediately.

  - o Ordinary nylon sleeves → no protection against oil injection →
  - o A special sleeving system is required → (LifeGuard™) from Gates.

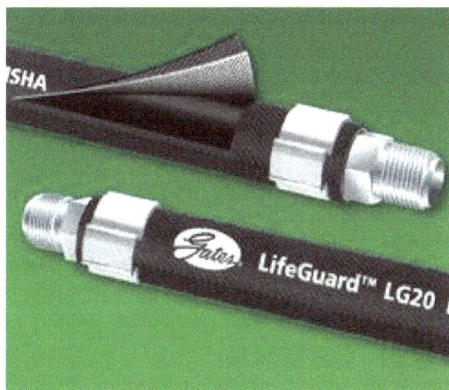

🎥 Video 283 (4.5 min)

**Fig. 3.86- Hose Oil-Injection
Protection Sleeve "Lifeguard"
(Courtesy of Gates)**

115

115

3.6.10- Hose Whip Restraints

- If a pressurized hose separating from its fitting → Hose Wipe Restrainer
  - prevent whipping of the hose →
  - Prevent damage to nearby equipment.
  - Prevent injury to nearby operators.
  - Additional level of safety.

- The system consists of a hose collar and a cable assembly.

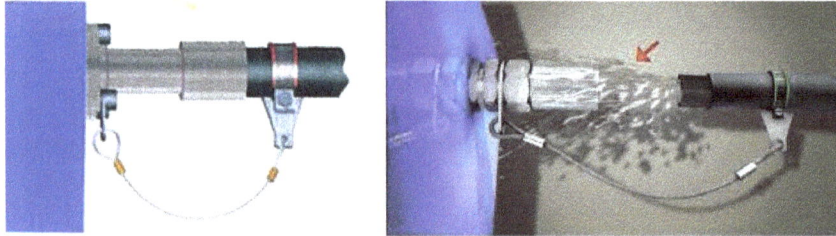

Fig. 3.95- Hose Wipe Restraint (Courtesy of Parker)

116

116

3.6.11- Hose Condition Monitoring

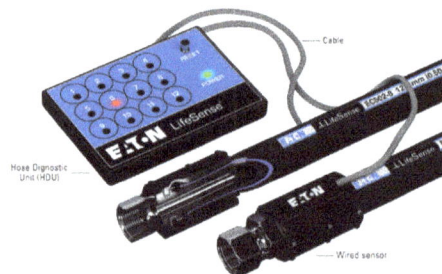

Fig. 3.96 - Real Time Hoses Condition Monitoring (www.hydrotechnik.com)

Built-In Reliability
Patented technology monitors hose condition in real time.

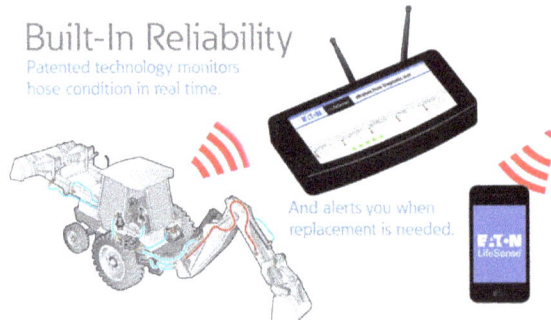

And alerts you when replacement is needed.

117

117

More Info about Hose Condition Monitoring review in textbook:

- **Figure 3.97**

118

3.6.12- Selection of Hose Couplings

- **Size:**
o Based on (size and type of the hose + port on hydraulic component).

- **Working Pressure:**
o Pmax for coupling should meets or exceeds Pmax for the hose.
o High pressures → O-ring-type fittings are recommended.
o Extremely high pressures → avoid use of swivel couplings.

- **Working Temperature:**
o Extreme T fluctuations → Mated surfaces fittings are recommended.
o High working temperature → avoid use of O-rings of sealing fittings.

- **Fluid Compatibility:**
o Hoses, couplings, O-rings, are affected by the fluid.
o Check fluid compatibility and chemical resistance charts.

119

- **Corrosion Resistance:**
  o Marked on the fittings.
  o Tested under SAE J516 and ASTMB I 17 Salt-Spray Tests.

- **Vibration:**
  o Motion and/or vibration→ Coupling can potentially weaken or loosen.
  o → Avoid use of threaded sealing coupling.
  o → Split flange or O-ring sealing fittings are recommended.

- **Use of Quick Connect Coupling:**
  o Check if any of the two ends are pressurized.
  o → If yes, use the ones with built in check valves.

Video 136 (0.5 min)

120

120

- **Use of Adaptors:**
  o Connect directly to a port → less # of fittings.
  o Connect to adapters → Ease connection and routing.

Fig. 3.98 - Use of Adaptors (Courtesy of Gates)

121

121

3.6.13- Hydraulic Hoses Selection Criteria

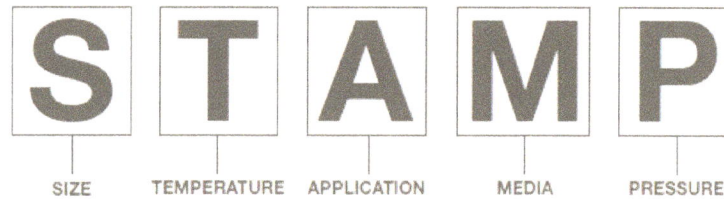

Fig. 3.99 - Keyword for Selecting Hydraulic Hoses (Courtesy of Parker)

- **Size:**
  o Actual internal area.
  o Flow pattern (Laminar/Turbulent)
  o Allowable pressure losses for suction lines.
  o → Larger size is considered

- **Temperature:**
  o A hose must meet or exceed the application working temperature.

122

122

- **Application:**
  o Considerations: rubbing, abrasive surfaces, rotations, etc.

- **Fluid Compatibility:**
  o Hose manufacturers should provide hose-fluid compatibility charts.

- **Pressure:**   Video 135 (3 min)
  o Static Pressure: System pressure distribution analysis is needed.
  o Dynamic Pressure: Pressure spikes, surges, peaks or overshoot)
  o Inside Negative Pressure: Suction hoses
  o Outside Positive Pressure: under water applications.
  o Low-Pressure Hoses: 20 bar (300 psi). App (lubrication & drain lines).
  o Medium-Pressure Hose: 20-200 bar (300 – 3000 psi). App (heavy-duty trucks, fleet vehicle, and aircrafts).
  o High-Pressure Hoses (2-Wire Braided): 400 bar (6000 psi). App (construction equipment).
  o Extremely High-Pressure Hoses (4-wire or 6-wire braded): >400 bar (>6000 psi). App (off-highway equipment and heavy-duty machinery with high impulse or pressure surges).

123

123

- **Electrical Conductivity:**
  - General Applications → hose/fitting interface to be sufficiently conductive to drain off static electricity.
  - Applications that need to maintain electrical isolation
  - → regular nonconductive hoses are used to prevent electrical current flow.
  - Nearby High Voltage Lines
  - → special nonconductive hoses are required.
- **Hose Length:**
  - Hose length must be wisely specified (neither short nor long).
  - Short hoses → overstretched when pressurized → detaching of end joints.
  - Long hoses → vibration, rubbing, pressure losses, etc.

124

- **Hoses Pictogram:**

Video 747

# Part Number	Standard Cover	Tough Cover	Super Tough	Hose I.D.		Hose O.D.		Working Pressure		Minimum Bend Radius		Weight		Vacuum Rating (Hg)	
	387	387TC	387ST												
387	ISO 18752 Performance			inch	mm	inch	mm	psi	MPa	inch	mm	lbs/ft	kg/m	inches of Hg	kPa

Fig. 3.100 - Hydraulic Hoses Pictogram (Courtesy of Parker)

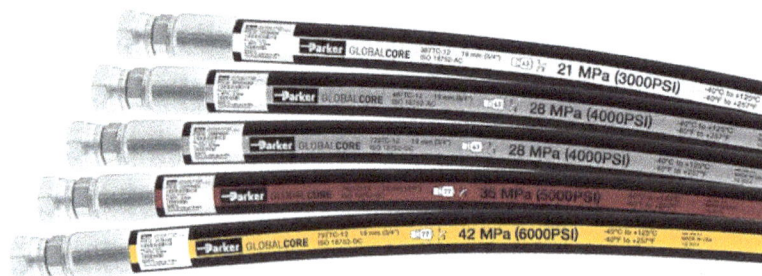

Fig. 3.101 - Example of Hydraulic Hoses with Pictogram Marked (Courtesy of Parker)

125

**Example of Hoses from Industry
review in textbook:**

- **Figure 3.103 through 3.107**

126

126

Chemical Name	Neoprene (Poly-Chloroprene)	Nitrile (Acrylonitrile and Butadiene)	Butyl (Isobutylene and Isoprene)	Hypalon (Chlorosulfonated Polyethylene)	EPDM (Ethylene Propylene Diene)	CPE (Chlorinated Polyethylene)
ASTM-SAE Designation SAE J14 & SAE J200	SC / BC	SB / BG	R / AA	TB / CE	R / AA	None / None
Flame Resistance	Very Good	Poor	Poor	Good	Poor	Good
Petroleum Base Oils	Good	Excellent	Poor	Good	Poor	Very Good
Diesel Fuel	Good to Excellent	Excellent	Poor	Poor	Poor	Very Good
Resistance to Gas Permeation	Good	Good	Outstanding	Good to Excellent	Fair to Good	Good
Weather	Good to Excellent	Poor	Excellent	Very Good	Excellent	Good
Ozone	Good to Excellent	Poor for Tube Good for Cover	Excellent	Very Good	Outstanding	Good
Heat	Good	Good	Excellent	Very Good	Excellent	Excellent
Low Temperature	Fair to Good	Poor to Fair	Very Good	Poor	Good to Excellent	Good
Water-Oil Emulsions	Excellent	Excellent	Good	Good	Poor	Excellent
Water-/Glycol Emulsions	Excellent	Excellent	Excellent	Excellent	Excellent	Excellent
Diesters	Poor	Poor	Excellent	Fair	Excellent	Very Good
Phosphate Esters	Fair (For Cover)	Poor	Good	Fair	Very Good	Very Good
Phosphate Ester Base Emulsions	Fair (For Cover)	Poor	Good	Fair	Very Good	Very Good

Table 3.15

Fluid Compatibility

127

127

135/335

3.7- Flanges for Transmission Line Connections

❖ **Flanges versus Fittings:**
▪ <u>Connection and Sealing:</u>
o Fittings → tightening threads between the mating halves.
o Flanges → bolting two mating halves that have an O-ring in between.

o <u>Line Size:</u>
o Fittings → effective and easy to install for smaller hoses (< 1" OD).
o Flanges → work best for larger size hoses (> 1" OD)
o Flanges → recommended for high pressures, vibration, and shock loads.
o Flanges → ideal for installation where there is not enough swing clearance for a wrench.

Video 684 (4 min)

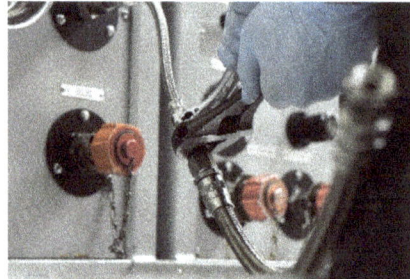

Fig. 3.108 - Use of fittings in Hoses (mac-hyd.com)

128

128

❖ **Flanges According to SAE-J518 Standard:**

▪ **SAE Standard Code 61** is for Pmax = (3000 – 5000) depending on the size.
▪ **SAE Standard Code 62** is for Pmax = 6000 psi regardless of size.
▪ Some OEM have their own flange dimensions.

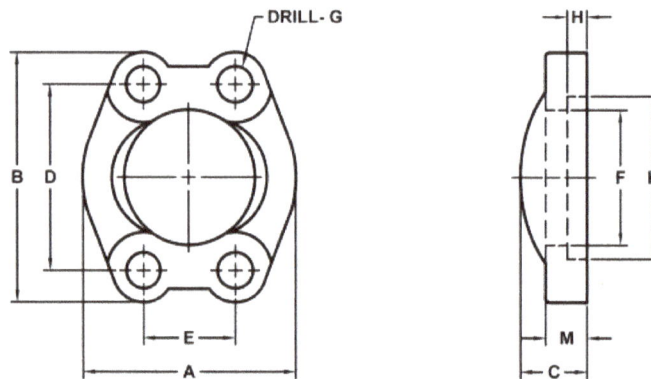

Fig. 3.109 - Standard Flange Dimensions

129

129

❖ **Components of Flange Assembly:**

Split Flange

One Piece Flange

Clamping Bolts

Code 61/62 Flange

Line End

Flange Head

O-ring Seal

Component Connection Port

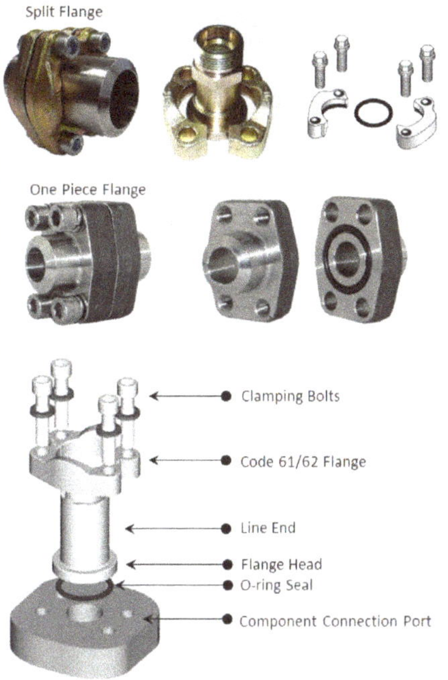

Fig. 3.110- Components of Flange Assembly

Video 662 (1 min)

❖ **One-Piece vs. Split Flanges:**

- Flanges are used with all kind of lines (Pipes, tubes, and hoses)

- One-piece welded-end flanges → connect pipe sizes up to 5 inches.

- Split flanges → more adequate for connecting tubes and hoses.

130

130

Fig. 3.111 - Hydraulic Pipe Flange Assembly

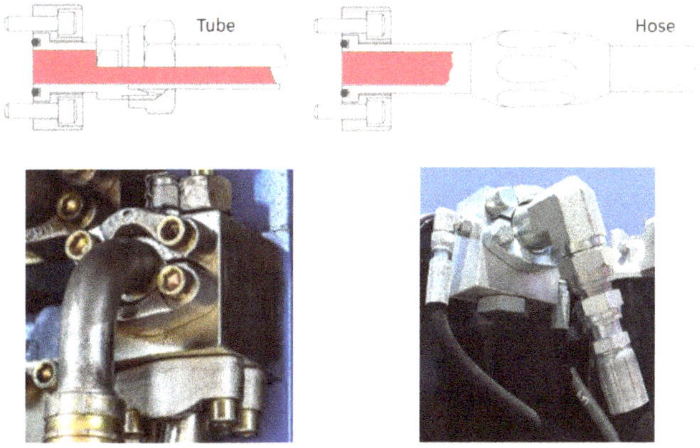

Tube   Hose

Fig. 3.104 - Hydraulic Tube and Hose Flange Assembly

131

131

❖ **Square Flanges:**

▪ Connect two tubes or two pipes.

▪ Connect a tube or a pipe to a hydraulic component.

▪ Various dimensions and materials (mainly carbon steel).

▪ Standards:
  o ANSI/ISO 6164/ASME Square flange.
  o JIS B2291, JIS F7806, DIN 3901, BS 16.5.

▪ Also customized to customers' specifications.

Fig. 3.113 - Square Flanges (https://flanges-pipe.com)

132

132

3.8- Rubber Expansion Fittings

❖ **Reasons to use Rubber Expansion Fittings:**

▪ Temperature fluctuation → thermal stresses → expansion and contraction.

▪ (Movement of foundations + load fluctuations + vibration during startup + flow surges) → mechanical stresses → expansion and contraction.

▪ Isolate structure-borne noise.

▪ Accommodating inaccuracy and misalignment in installation.

Fig. 3.114 - Rubber Expansion Fittings (www.grainger.com)

133

133

❖ **Material for Rubber Expansion Fittings:**

▪ Natural or synthetic rubber depending on the pressure.

❖ **Working Conditions for Rubber Expansion Fittings:**

▪ Pressures: <20 bar (300 psi) as in suction and drain lines.

▪ Temperature: up to +10°C.

134

134

3.9- Test Points

❖ **Function of Test Points:**
▪ Pressure monitoring and venting.
▪ Sampling of hydraulic fluids.
▪ Connection with instruments and sensors.

❖ **Features of Test Points:**
▪ Ensures 100% leakage free coupling while under system pressure.
▪ Available sealing materials are NRB, FKM and EPDM.
▪ Available styles and sizes.
▪ Available standards are ISO, SAE, and international threads
▪ Available in steel and stainless steel.

Fig. 3.115 - Series 1620 Test Points
(www.hydrotechnik.com)

135

135

❖ **Innovative Test Point:**

▪ An innovative test points (ICON™).

▪ Easy and tight seal connection with various fluid power components.

▪ Connect to various tube ends configurations.

▪ Sensing capabilities.

▪ Instant LED visual feedback.

▪ Data output to verify sealing connection accuracy.

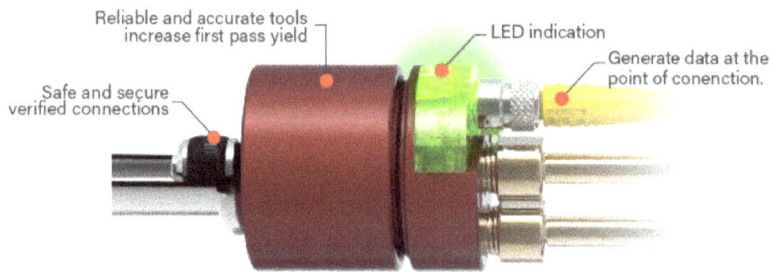

Fig. 3.116 - Innovative Test Points (Courtesy of CEJN)   136

136

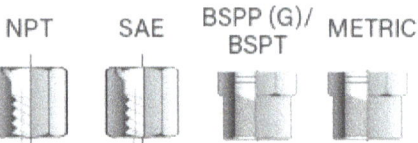

Fig. 3.116 - Continue   137

137

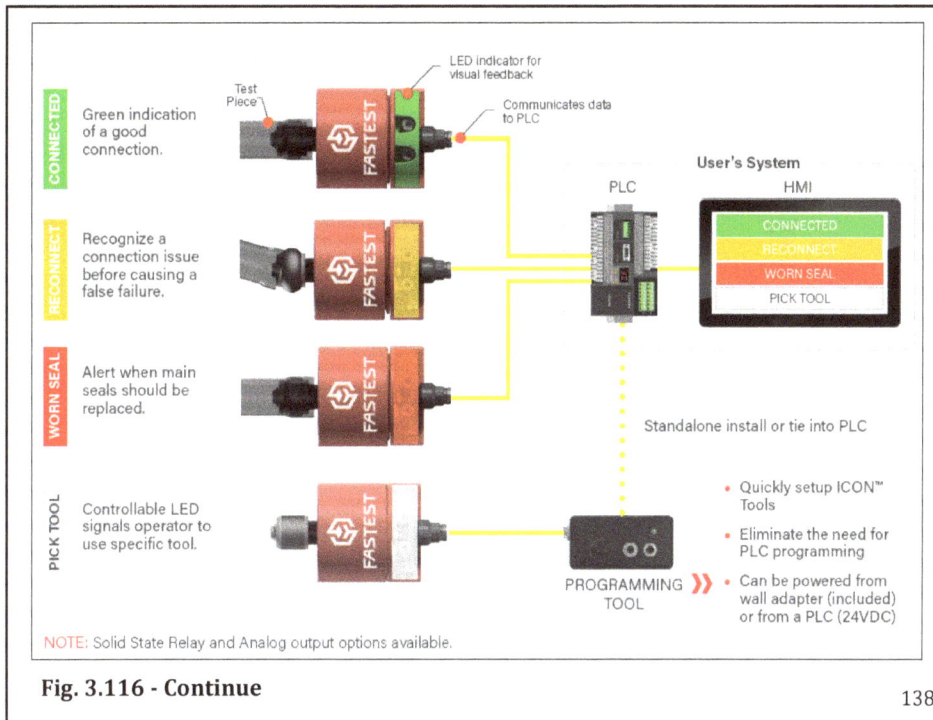

Fig. 3.116 - Continue

138

138

3.10- Pressure Measurement Hoses

❖ **Function of Measurement Hoses:**

▪ Pressure measuring.

▪ Transmitting control pressure signals.

▪ System diagnostic.

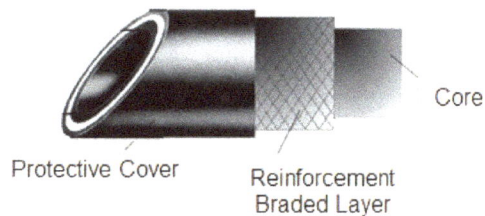

Fig. 3.117 - Construction of Measurement Hoses (www.hydrotechnik.com)

139

139

❖ **Features of Measurement Hoses**:
- Measure pressures of up to 600 bar (9000 psi).
- Resistance to aggressive fluids.
- Supported by anti-kinking spirals or aluminum protective covers.

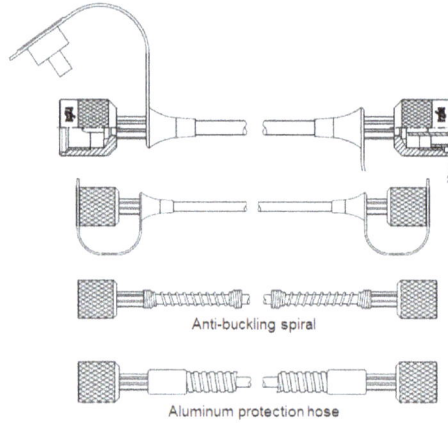

Anti-buckling spiral

Aluminum protection hose

Fig. 3.118 - Measurement Hoses Supported by Anti-Buckling and Protection Covers (www.hydrotechnik.com)

140

140

- Can be combined with test points.

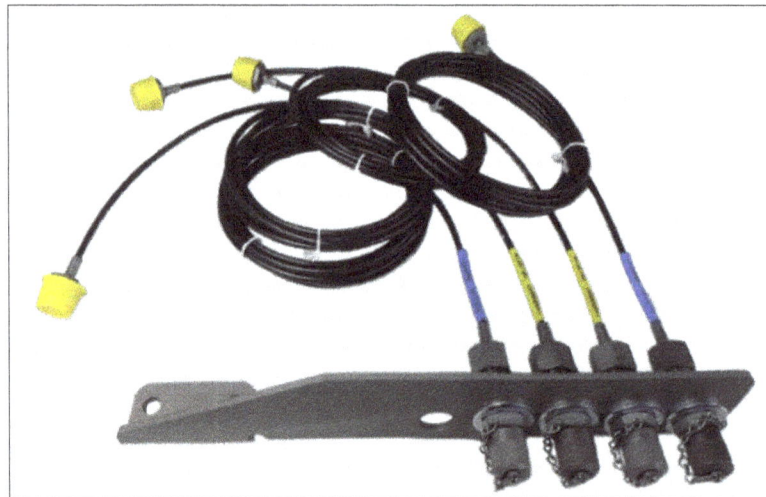

Fig. 3.119 - Measurement Hoses Combined with Test Points (www.hydrotechnik.com)

141

141

- Pressure drop should be reported by the manufacturer.

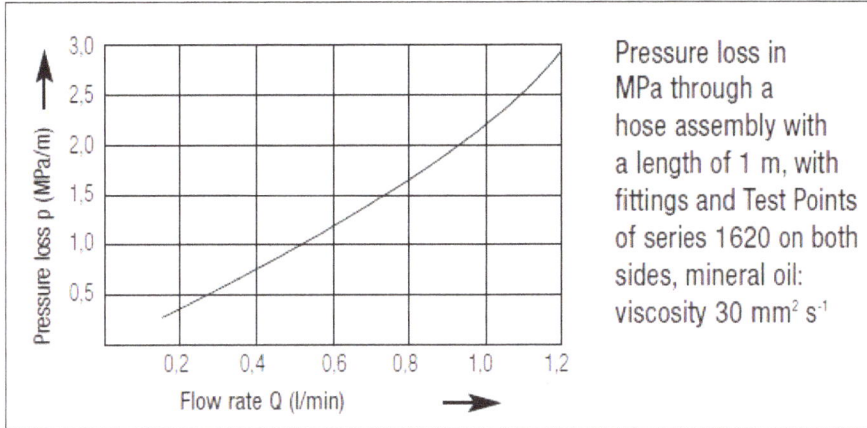

Pressure loss in MPa through a hose assembly with a length of 1 m, with fittings and Test Points of series 1620 on both sides, mineral oil: viscosity 30 mm^2 s^{-1}

Fig. 3.120 - Pressure Drop in Measurement Hoses
(www.hydrotechnik.com)

142

142

❖ **Routing of Measurement Hoses:**

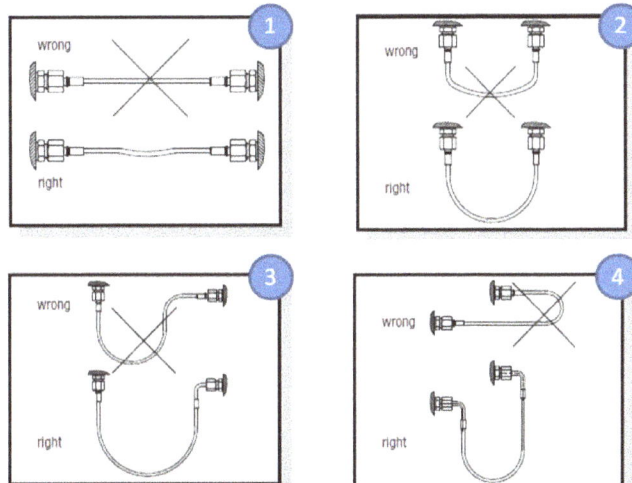

Fig. 3.121 - Best Practices of Routing Measurement Hoses
(www.hydrotechnik.com)

143

143

3.11 Manifolds

- Manifolds are blocks machined with internal passages →
- Compact, fast response, and energy efficient systems.
- Standards: ANSI/NFPA.
- Software → Design manifold based on hydraulic circuit diagram.

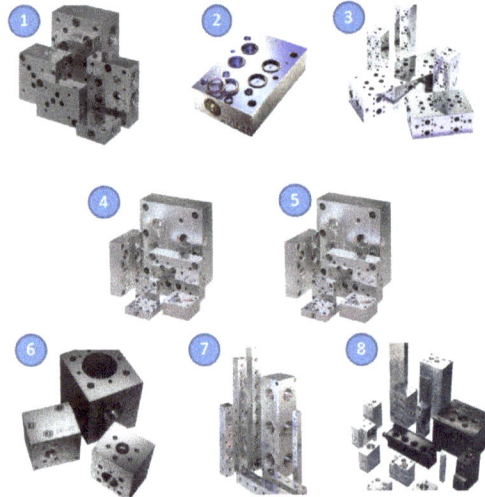

1. Subplates
2. Cover Plates
3. Bar Manifolds
4. Tapping Plates
5. Valve Adaptors
6. Din Bodies
7. Header & Junction Blocks
8. Custom Engineered Manifolds

Fig. 3.122 – Hydraulic Manifolds
(www.daman.com)

144

Chapter 3 Reviews

1. If the allowable working pressure for a conductor is 1000 psi with the consideration of a safety factor equal to 4, what is the burst pressure for that conductor?
    A. 250 psi.
    B. 500 psi.
    C. 2000 psi.
    D. 4000 psi.

2. Nominal size of pipes and tubes is rated by?
    A. Inner diameter (ID).
    B. Outer diameter (OD).
    C. Wall Thickness.
    D. None of the above.

3. Nominal size of flexible hoses is rated by?
    A. Inner diameter (ID).
    B. Outer diameter (OD).
    C. Wall Thickness.
    D. None of the above.

4. A One-foot pipe selected from schedule 80 is?
    A. Heavier than same length of a pipe selected from schedule 160.
    B. Has less maximum pressure rating than a pipe selected from schedule 40.
    C. Has larger inner diameter than a pipe selected from schedule 40.
    D. None of the above.

5. Pressure hose can't replace suction hose because?
    A. Inner layers of a pressure hose are not attached to reinforcing wires so that, if a pressure hose is subjected to a negative gauge pressure, inner layers may collapse and block the hose.
    B. Pressure hose has smaller inner diameter than suction hose.
    C. Suction hose has thicker walls than pressure hose.
    D. Suction hose has larger outer diameter than pressure hose.

6. A hose of (-8) size means?
    A. A hose of 8 inches outer diameter.
    B. A hose of 8 inches inner diameter.
    C. A hose of a half inch inner diameter.
    D. A hose of a half inch outer diameter.

Chapter 3 Assignment

Student Name: --- Student ID: ------------------

Date: --- Score: ------------------------

Assignments: Use the shown below chart to find the range of the intake pipe diameter for a pump discharge = 100 GPM

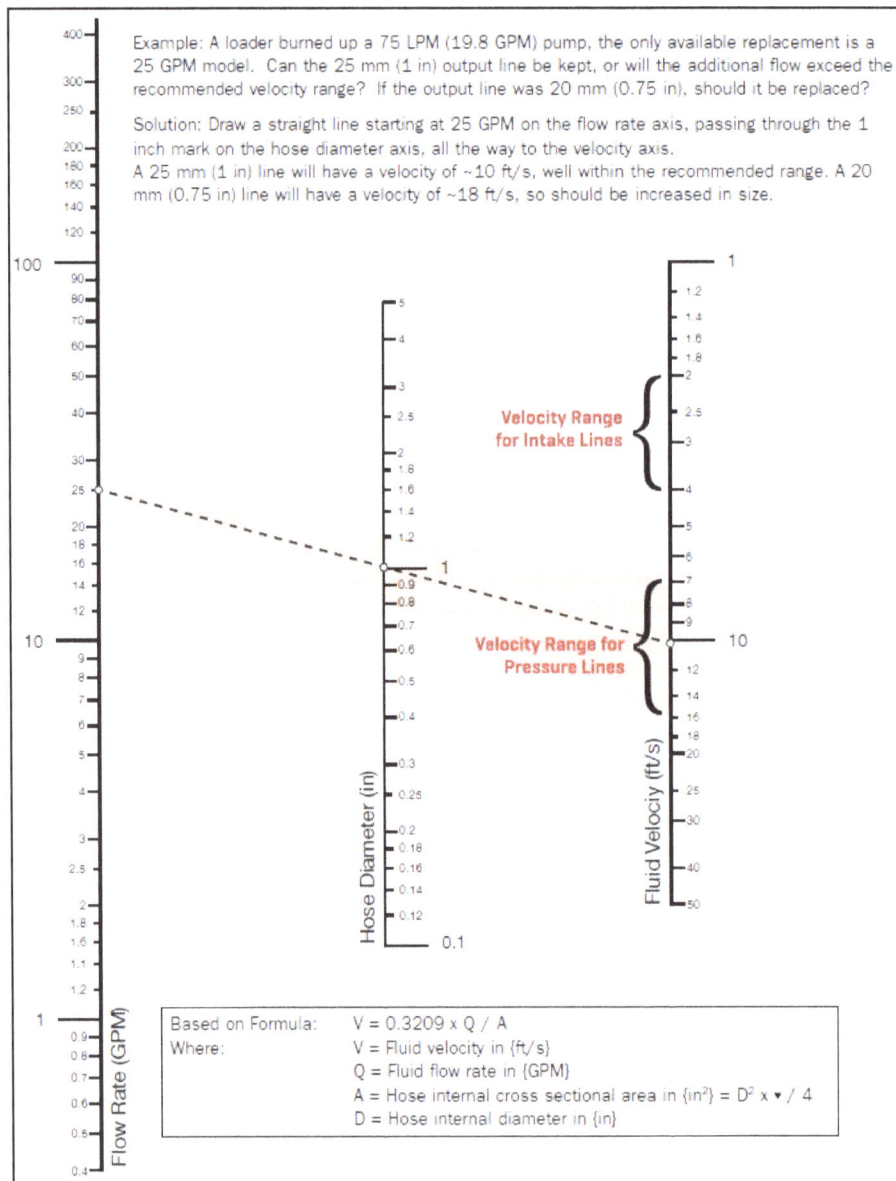

Chapter 4
Hydraulic Sealing Elements

Objectives:

This chapter provides a knowledge base for fluid power users to become familiar with the commonly used seals in hydraulic components. This chapter presents an overview of hydraulic sealing elements including seal functions, classifications, and materials. This chapter also presents 15 various properties of hydraulic seals and the relevant standard test methods. This chapter also presents sealing solutions for cylinders and rotational shafts.

0

0

Brief Contents:

4.1- Introduction to Hydraulic Sealing Elements
4.2- Sealing Rings
4.3- Cup Seals
4.4- U-Cup Seals
4.5- T-Shaped Seals
4.6- V-Packings
4.7- Spring-Energized Seals
4.8- Wear-Rings
4.9- Backup Rings
4.10- Rod Wipers
4.11- Materials for Hydraulic Sealing Elements
4.12- Properties and Test Methods for Hydraulic Sealing Elements
4.13- Best Practices for Hydraulic Seals Selection
4.14- Sealing Solutions for Hydraulic Cylinders
4.15- Sealing Solutions for Rotational Shafts

1

1

4.1- Introduction to Hydraulic Sealing Elements

4.1.1-Functions of Hydraulic Sealing Elements

❖ Hydraulic seals are basically used to:

- Prevent external leakage →
  - Cost saving.
  - Environment. al damage.
  - Safety.
  - Liability.

- Prevent internal leakage →
  - Pressure losses.
  - Operation efficiently.
  - Heat generation.

- Provide controlled lubrication for adjacent parts or surfaces.

- Other functions →
  - Remove dirt, dust, and other contaminants from getting into the hydraulic components.

2

2

Leakage Rate	Monthly Losses	Yearly Losses
1 Drop/5Sec.	6.6 Gallon = $66	80 Gallon = $800
1 Drop/Sec.	34 Gallon = $340	409 Gallon = $4090
3 Drop/Sec.	113 Gallon = $1130	1243 Gallon = $12430
Steady Stream	720 Gallon = $7200	8640 Gallon = $86400

Video 256 (2.0 min)

Spend tens of dollars to change the seal on time

Save hundreds of dollars on Troubleshooting

Save thousands of dollars on machine shutdown

Fig. 4.1 - Cost of Oil Leakage

3

3

4.1.2-Applications of Hydraulic Sealing Elements

Video 257 (1.5 min)

Fig. 4.2 - Examples of Sealing Applications (Courtesy of Trelleborg)

4

4

4.1.3-Classifications of Hydraulic Sealing Elements

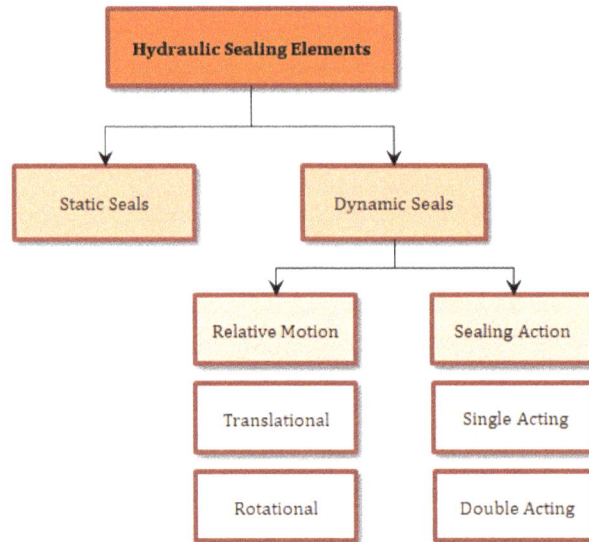

Fig. 4.3 - Classifications of Hydraulic Sealing Elements

5

5

❖ **Static Seals:**

- No (motion or relative motion).

- Used to fill confined or none-confined spaces.

- Basic static seals include Sealing Rings and Gaskets.

❖ **Dynamic Seals:**

- Installed on a stationary surface, e.g. Cylinder rod seal.

- Installed on a moving surface, e.g. Cylinder piston seal.

- Translational (as in cylinders and spool valves)

- Rotational (as in pumps and motors).

- Seal in one or two directions.

- Lubricate between surfaces.

6

6

❖ **Common Sealing Configurations:**

- Sealing Rings

- Cup Seals

- U-Cup Seals

- T-Shaped Seals

- V-Packings

- Spring-Energized Seals

- Glands

- Wear-Rings

- Back-up Rings

- Rod Wipers (Scrapers)   Video 254 (10 min)

7

7

4.2- Sealing Rings

4.2.1- Features and Basic Use of Sealing Rings

❖ **Advantages:**

- They are inexpensive and readily available.

- Used for both static and dynamic sealing

- Used for both translational and rotational shafts.

- They are available in all sizes and all type of elastomers.

- They work over a wide range of operating pressure.

- They are easy to assemble and to replace.

- They require small room to fit inside the components.

- Their failure analysis isn't difficult.

- Limited use in high temperature application.

8

8

- Sold as separate pieces with specific dimensions or as *Cord Stock*.

Fig. 4.4 - Example of O-Rings Cord Stock
(Courtesy of MFP Seals)

9

9

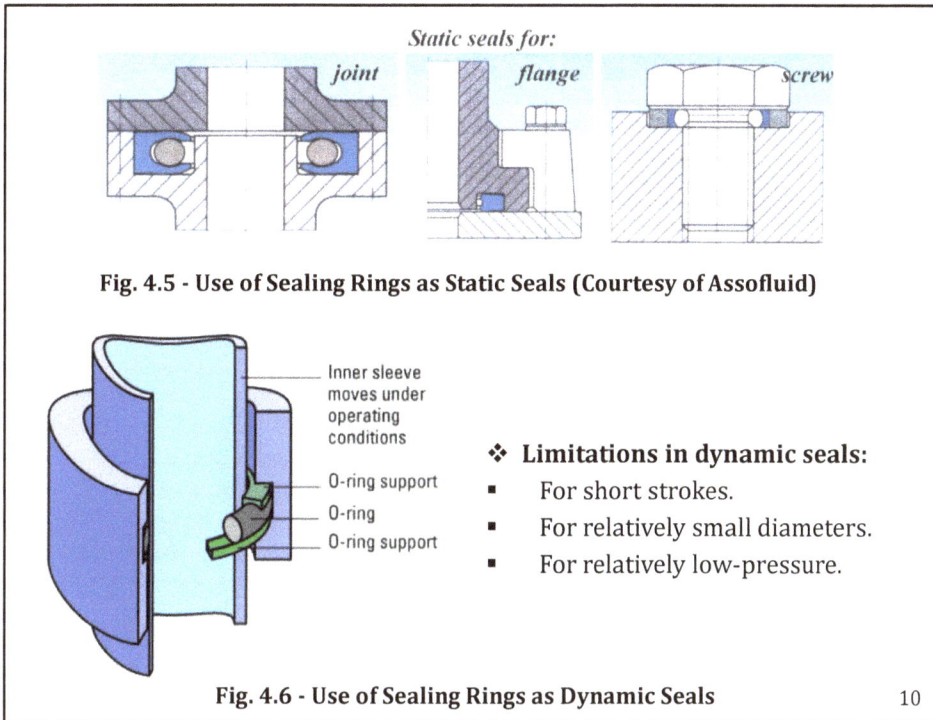

Fig. 4.5 - Use of Sealing Rings as Static Seals (Courtesy of Assofluid)

Inner sleeve moves under operating conditions

O-ring support

O-ring

O-ring support

❖ **Limitations in dynamic seals:**
- For short strokes.
- For relatively small diameters.
- For relatively low-pressure.

Fig. 4.6 - Use of Sealing Rings as Dynamic Seals   10

10

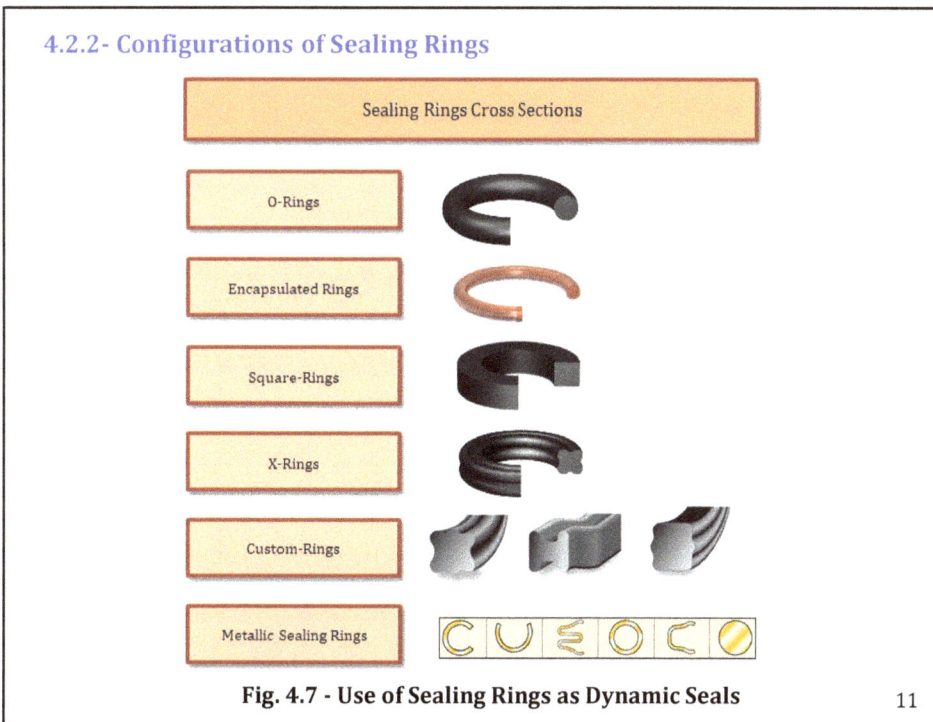

4.2.2- Configurations of Sealing Rings

Sealing Rings Cross Sections

O-Rings

Encapsulated Rings

Square-Rings

X-Rings

Custom-Rings

Metallic Sealing Rings

Fig. 4.7 - Use of Sealing Rings as Dynamic Seals   11

11

4.2.3- O- Rings

- Most commonly known type of sealing rings.

4.2.3.1- O-Rings Construction

- O-Ring → solid circular cross section.
- Available in a wide range of materials and dimensions.

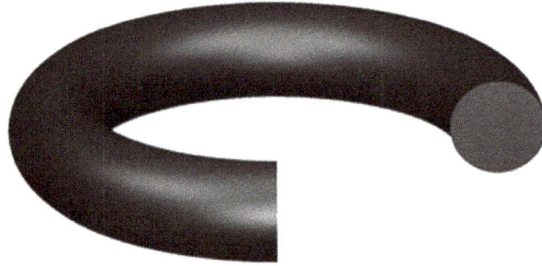

Fig. 4.8 - Basic Construction of Static O-Rings (Courtesy of MFP Seals)

12

12

4.2.3.2- O-Rings Sealing Mechanism

- After assembly → O-Ring is compressed (15-30) % of its initial volume.
- The initial squeeze in a radial or axial direction → initial sealing condition.
- Pressure applied → sealing condition α system pressure.

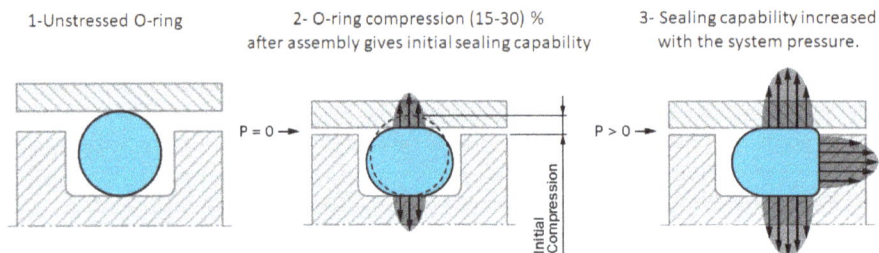

Fig. 4.9 - O-Ring Sealing Mechanism (Courtesy of Trelleborg)

Video 014 (0.1 min)

13

13

4.2.3.3- O-Rings Main Dimensions

- O-Rings are produced in inch and metric standard sizes.
- **Standards:** ISO 3601 and SAE AS568.
- Characterizing Dimensions:
  - Inside Diameter (ID).
  - Radial and Axial Crosssection Width (W).
  - Radial and axial Flashes due to rubber injection.

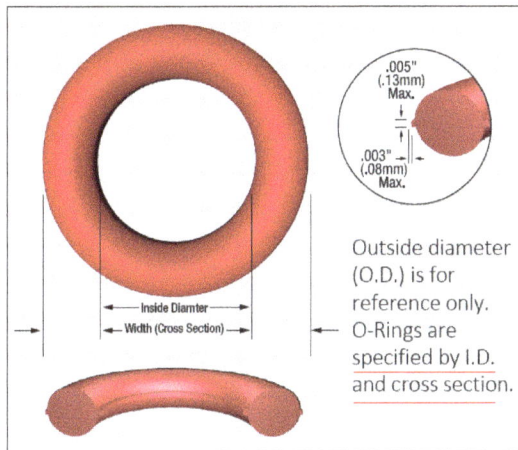

.005" (.13mm) Max.

.003" (.08mm) Max.

Outside diameter (O.D.) is for reference only. O-Rings are specified by I.D. and cross section.

Inside Diamter
Width (Cross Section)

**Fig. 4.10 - Main Dimensions
of Basic O-Rings
(applerubber.com)**

14

14

4.2.4- Encapsulated O-Rings

- An elastomeric base material encapsulated with Teflon coating (FEP) →
- Increased chemical, temperature and wear resistance.

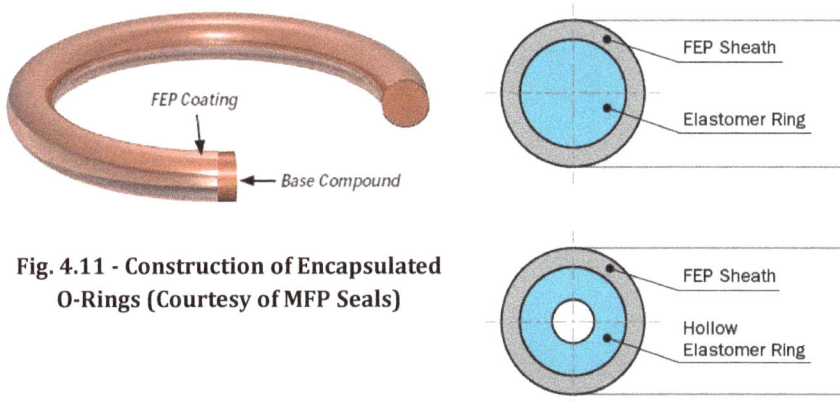

FEP Coating

Base Compound

**Fig. 4.11 - Construction of Encapsulated
O-Rings (Courtesy of MFP Seals)**

FEP Sheath

Elastomer Ring

FEP Sheath

Hollow
Elastomer Ring

**Fig. 4.12 - Typical Construction of Encapsulated
O-Rings (Courtesy of Trelleborg)**

15

15

4.2.5- Square-Rings

- Interchangeable with the O-Rings (in certain sizes).
- Larger sealing surface → handle higher pressures than O-Rings.
- Commonly used for static applications.

Fig. 4.13 - Construction of Square-Rings (Courtesy of MFP Seals)

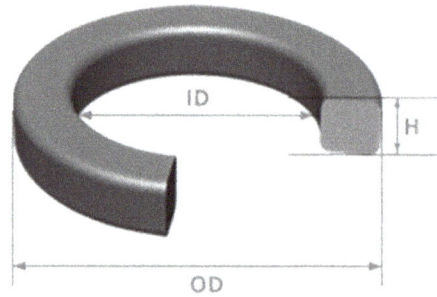

Fig. 4.14 - Main Dimensions of Square-Rings (www.marcorubber.com)

16

16

4.2.6- X- Rings

- X-Rings, also referred to as Quad-Rings.
- Dual sealing lip (2 contact points vs. 1 contact point in O-Ring)→
  - Better sealing with less squeeze.
  - Reduced friction → Less seal wear → longer service life.
  - Resists twisting in the groove → good for rotary applications.

Fig. 4.15 - Construction of X-Rings (Courtesy of MFP Seals)

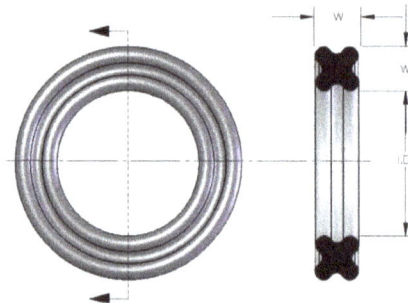

Fig. 4.16 - Main Dimensions of X-Rings (www.marcorubber.com)

17

17

4.2.7- Custom-Rings

- **Example 1:**
  o Dynamic sealing applications.
  o Excellent sealing → Near zero leakage at P up to 138 bar (2000 psi).
  o Used with standard O-Ring grooves.

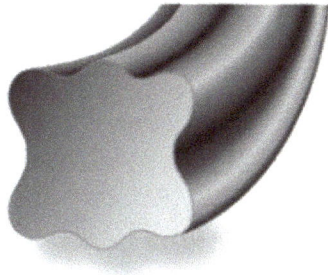

Fig. 4.17 - Quad-O-Dyn® Brand Seals
(www.mnrubber.com)

18

18

- **Example 2:**
  o Static face seal applications.
  o Ideal to fill single or multiple grooves.

Fig. 4.18 - Quad®-O-Stat Brand Seals
(www.mnrubber.com)

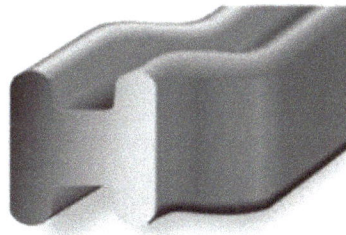

- **Example 3:**
  o Static face sealing applications.
  o Each of the six contact points serves as an individual seal.
  o If one lobe fails, the remaining lobes provide zero leakage sealing.
  o They can be installed in standard O-Ring grooves.

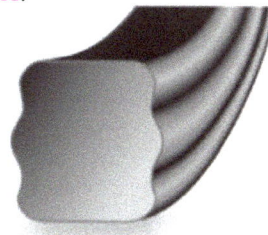

Fig. 4.19 - Quad® P.E. Plus Brand Seals
(www.mnrubber.com)

19

19

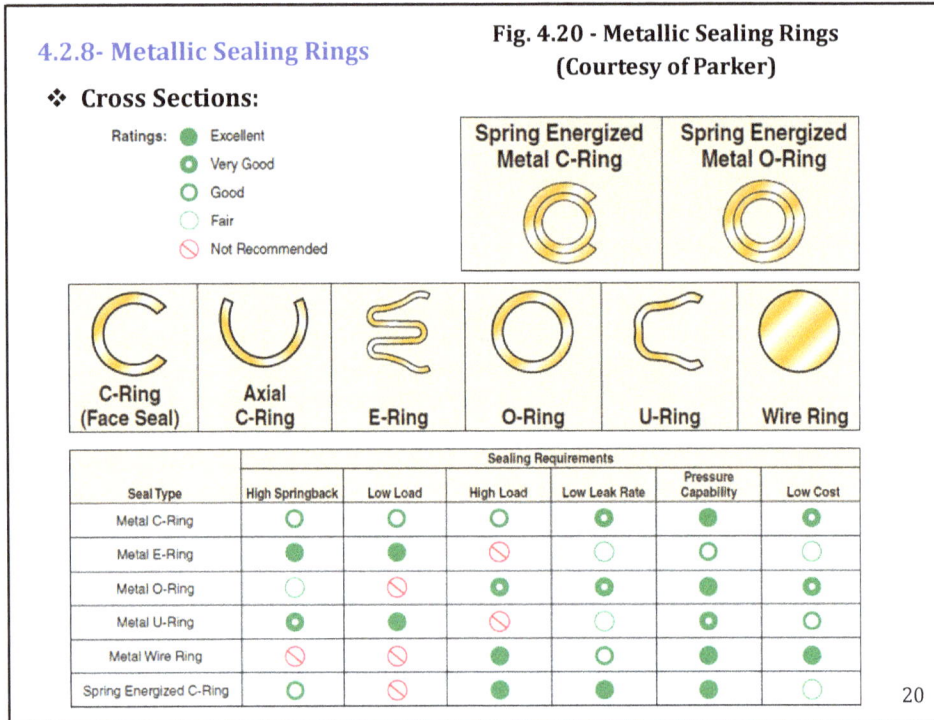

4.2.8- Metallic Sealing Rings

Fig. 4.20 - Metallic Sealing Rings (Courtesy of Parker)

❖ **Cross Sections:**

Ratings:
- ● Excellent
- ◉ Very Good
- ○ Good
- ◌ Fair
- ⊘ Not Recommended

Spring Energized Metal C-Ring | Spring Energized Metal O-Ring

C-Ring (Face Seal) | Axial C-Ring | E-Ring | O-Ring | U-Ring | Wire Ring

Seal Type	High Springback	Low Load	High Load	Low Leak Rate	Pressure Capability	Low Cost
Metal C-Ring	○	○	○	◉	●	◉
Metal E-Ring	●	●	⊘	○	○	◌
Metal O-Ring	◌	⊘	◉	◉	●	◉
Metal U-Ring	◉	●	⊘	○	◉	○
Metal Wire Ring	⊘	⊘	●	○	●	●
Spring Energized C-Ring	○	⊘	●	●	●	◌

20

❖ **Material:**
- Various material (Steel, Stainless Steel, Cast Iron, Ductile Iron and Bronze).
- Piston Ring Coatings → Improve performance (lubricating requirements, corrosion resistance, wear, and friction reduction).
- Coating Material: Manganese Phosphate, Tin Nickel, PTFE/Nickel, Chrome and Silver)

❖ **Fluid Compatibility:** They are compatible with petroleum base and synthetic fluids and phosphate esters among others.

❖ **Major Advantages:**
- Reliable, severe operating conditions, and extended service life.
- High P up to 1700 bar (25,000 psi) without the risk of blow-by.
- High T up to 1800 °F (982 °C).

1. No inflow into the groove, and O-Ring is not activated by pressure
2. Fluid pressure rises rapidly pressing the piston seal down
3. O-Ring is compressed and the seal is leaking

Cylinder Wall
Piston Seal
O-Ring
Piston Head

What is Blow-By?

21

4.3- Cup Seals

- Basic Cup Seal shape.
- Sealing function in one direction.
- Applying pressure $\uparrow \rightarrow$ sealing force $\uparrow$.
- This cup seal requires a backing plate to retain it in position.
- Used for pressure up to 70 bar (1000 psi).

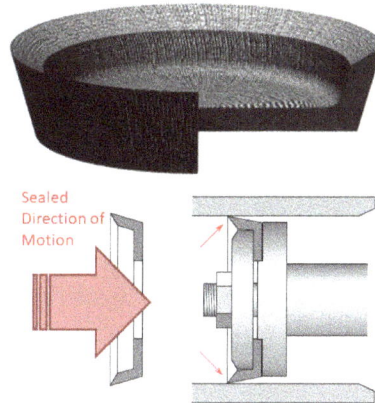

Sealed Direction of Motion

**Fig. 4.21 - Cup Seals
(Curtesy of MFP Seals)**

22

22

4.4- U- Cup Seals

Video 015 (0.2 min)

Video 320 (2.5 min)

- **Basic U-Cup Seals**
  - Sealing function in one direction
  - Static sealing lip and dynamic sealing lip.
  - Very popular as piston or rod seals in industrial cylinders.
  - Without backup ring → pressure up to 173 bar (2,500 psi).
  - Otherwise → seals extrusion could occur.

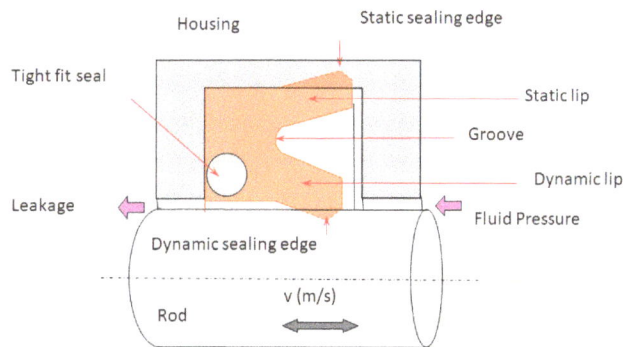

Housing
Static sealing edge
Tight fit seal
Static lip
Groove
Dynamic lip
Leakage
Fluid Pressure
Dynamic sealing edge
v (m/s)
Rod

Fig. 4.23 - U-Cup Sealing Mechanism

23

23

- **Symmetric O-Ring loaded U-Cup Seals:**
  - o Sealing function in one direction.
  - o Symmetric O-Ring loaded U-Cup seal.
  - o With and without anti-extrusion Back-up Ring.
  - o Working pressure 400 bar (5,800 psi).

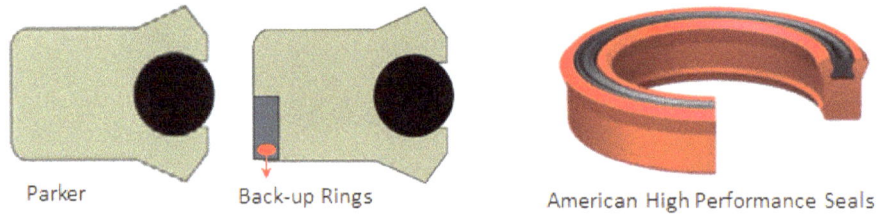

Parker   Back-up Rings   American High Performance Seals

Fig. 4.24 - Symmetric U-Cup Seals

24

24

4.5- T- Shaped Seals

- T-Shaped rubber seal + Two anti-extrusion Back-up Rings.
- Sealing function in two direction
- They are used for sealing both cylinder rods and pistons.
- Pressure up to 700 bar (10152 psi).

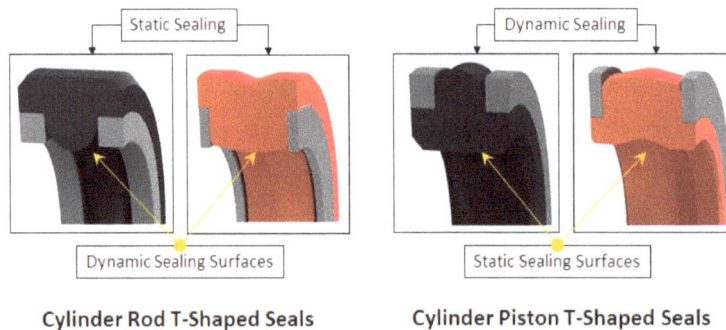

Static Sealing   Dynamic Sealing

Dynamic Sealing Surfaces   Static Sealing Surfaces

Cylinder Rod T-Shaped Seals   Cylinder Piston T-Shaped Seals

Fig. 4.25 -T-Shaped Seals (Courtesy of American High-Performance Seals)

25

25

- Supporting Back-up Rings on both sides →
- Eliminates rolling or spiraling
- Required for in (long stroke cylinders or dry rod conditions).

Fig. 4.26 - T-Shaped Sealing Package (Courtesy of MFP Seals)

26

26

4.6- V-Packings

- Multi-part sealing set.
- Used for sealing both cylinder rods and pistons.
- One (top) female adaptor (known as the Backup Ring or Base Ring).
- One(bottom) male adaptor (known as Compression or Energizing Ring).
- V-Rings made from elastomer with good extrusion resistance.

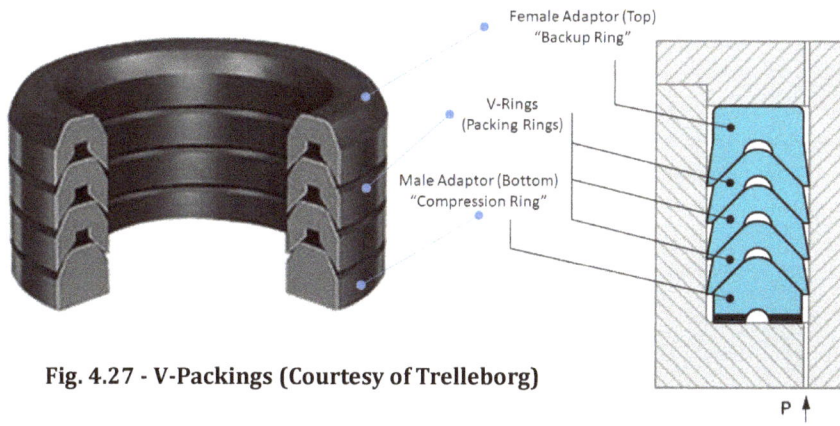

Female Adaptor (Top)
"Backup Ring"

V-Rings
(Packing Rings)

Male Adaptor (Bottom)
"Compression Ring"

Fig. 4.27 - V-Packings (Courtesy of Trelleborg)

P

27

27

- Different material in the same pack → best performance.

- Operating pressure ↑ → Number of V-Rings ↑

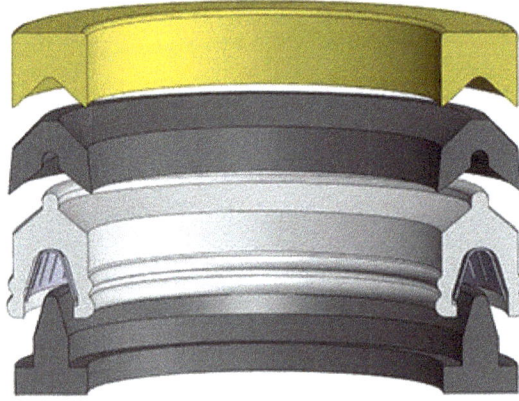

Fig. 4.28 - V-Packings of Different Materials
(www.hydrapakseals.com)

28

28

4.7- Spring-Energized Seals

Video 318 (2.5 min)

Video 319 (2.5 min)

- When Elastomeric seals cannot meet the application requirements (frictional, temperature, pressure, or chemical-resistance)

- → Spring-Energized Seals

- More dynamically stable.

- Common uses:

o For Static and dynamic sealing.

o With extreme operating T/P.

o With high surface speeds.

o With Non-lubricated surfaces.

o Explosive decompression resistant applications.

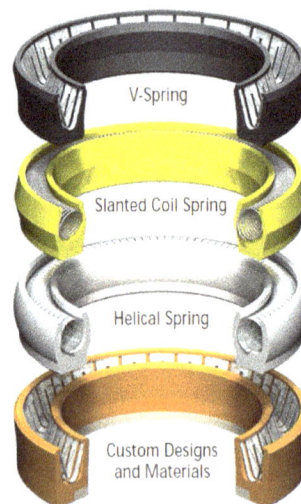

V-Spring

Slanted Coil Spring

Helical Spring

Custom Designs
and Materials

Fig. 4.29 - Spring-Energized Seals
(Courtesy of MFP Seals)

29

29

- Typical example for Spring-Energized Plastic U-Cup seals:
  - Suitable for reciprocating and rotary applications.
  - Low coefficient of friction.
  - Stick-slip free operation.
  - High abrasion resistance.
  - Dimensionally stable.
  - Resistant to most fluids, chemicals and gases.
  - Withstands rapid changes in temperature.
  - Excellent resistance to aging.
  - Interchangeable with O-Ring and Back-up Ring combinations.

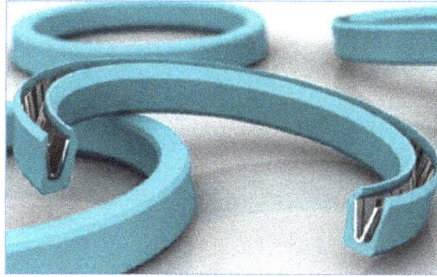

OPERATING CONDITIONS

Pressure:	Maximum dynamic load: 20 MPa Maximum static load: 40 MPa (207 MPa with back-up ring)
Speed:	Reciprocating up to 15 m/s Rotating up to 1.27 m/s
Operating temperature:	-70 °C to +300 °C Special Turcon and Zurcon® materials as well as alternative spring materials are available for applications outside this temperature range.
Media compatibility:	Virtually all fluids, chemicals and gases

Fig. 4.30 - Spring-Energized U-Cup Seals (Courtesy of Trelleborg)

30

30

- Sealing Mechanism:
  - Spring load → sealing at low pressures.
  - Applying P → sealing forces rises proportionally.

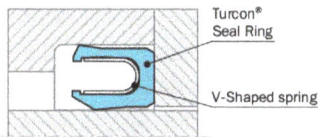

Turcon® Seal Ring

V-Shaped spring

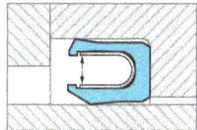

Spring Force without system pressure

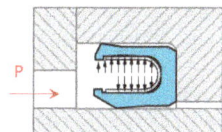

Sealing Force after applying system pressure

P

Fig. 4.31 - Sealing Mechanism of a Typical Spring-Energized Seal (Courtesy of Trelleborg)

31

31

4.8- Wear-Rings

- Hydraulic seals → do not provide bearing surfaces or carry lateral loads.
- Wear-Rings (also known as "Guide-Rings" or "Wear-Bands") → :
  o Used for cylinder rods and pistons.
  o Provide bearing surfaces to absorb side loads.
  o Prevent metal-to-metal contact → Leakage and component damage ↓.

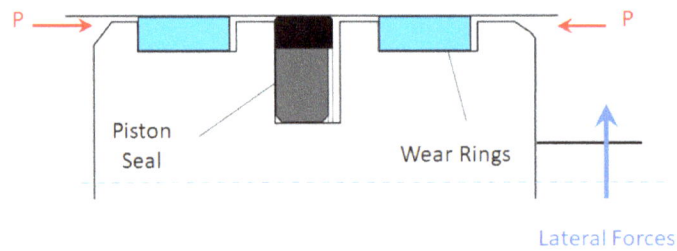

Fig. 4.32 - Wear-Rings

32

32

- Made of different material (Nylon, Teflon and other plastics).
- Available as pieces or strips.

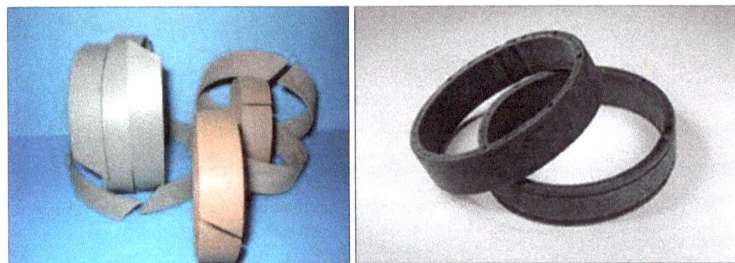

Fig. 4.33 - Examples of Wear Rings

Fig. 4.34 - Wear-Strip (Courtesy of MFP Seals)

33

33

- Available with various cut designs.
- Design with a special texture (named TEARDROP profile)
- → Small lubricant pockets on the surface
- → Improve the initial lubrication.

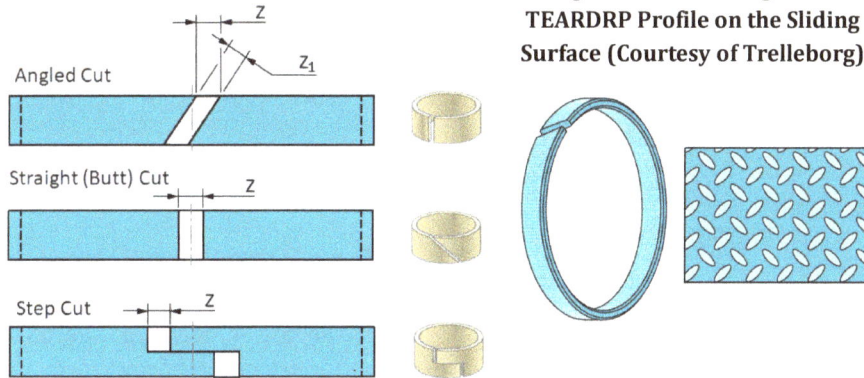

Fig. 4.36 - Wear-Rings with TEARDRP Profile on the Sliding Surface (Courtesy of Trelleborg)

Angled Cut

Straight (Butt) Cut

Step Cut

Fig. 4.35 - Wear-Rings Cut Design (Courtesy of Trelleborg)

34

- <u>Minimum Bearing Length.</u>

$$P_R = \frac{F \times f}{Projected\ Area} = \frac{F \times f}{D \times T} \rightarrow T = \frac{F \times f}{D \times P_R} \qquad 4.1$$

- P_R = Design Pressure for a specific Wear-Ring.
- F = maximum estimated lateral force.
- f = Safety factor (default = 2).
- D = Diameter of cylinder rod or piston.
- T = minimum bearing length.

- **Example:** A wear ring is required for a cylinder rod of 60 mm diameter and 40,000 N maximum estimated radial force. A specific Wear-Ring material is selected that has 100 N/mm² design pressure. Then the minimum width is:

$$T\ (mm) = \frac{40.000 \times 2}{60 \times 100} = 13.3$$

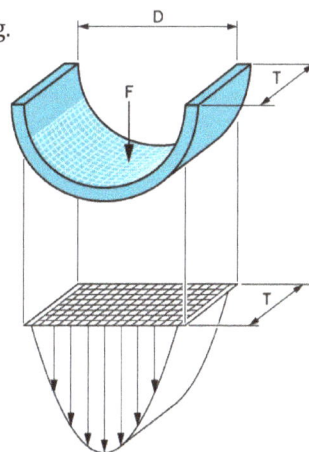

Fig. 4.37 - Calculation of the Bearing Length (Courtesy of Trelleborg)

35

Recommended Characteristics of Wear-Rings:
- High bearing load capabilities.
- High operating T & P.
- Cost effective.
- Easy installation and replacement.
- Wear-resistant and long service life.
- Low friction and self-lubrication.
- Wiping/cleaning effect.
- Ability to embed foreign particles possible.
- Damping of mechanical vibrations.

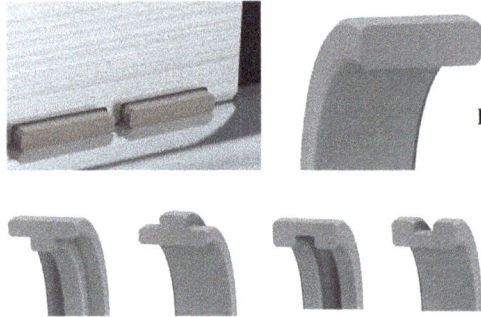

Fig. 4.38 - Wear-Rings of Various Cross Sections

(Courtesy of American High-Performance Seals)

36

36

Piston Head

Rod Gland

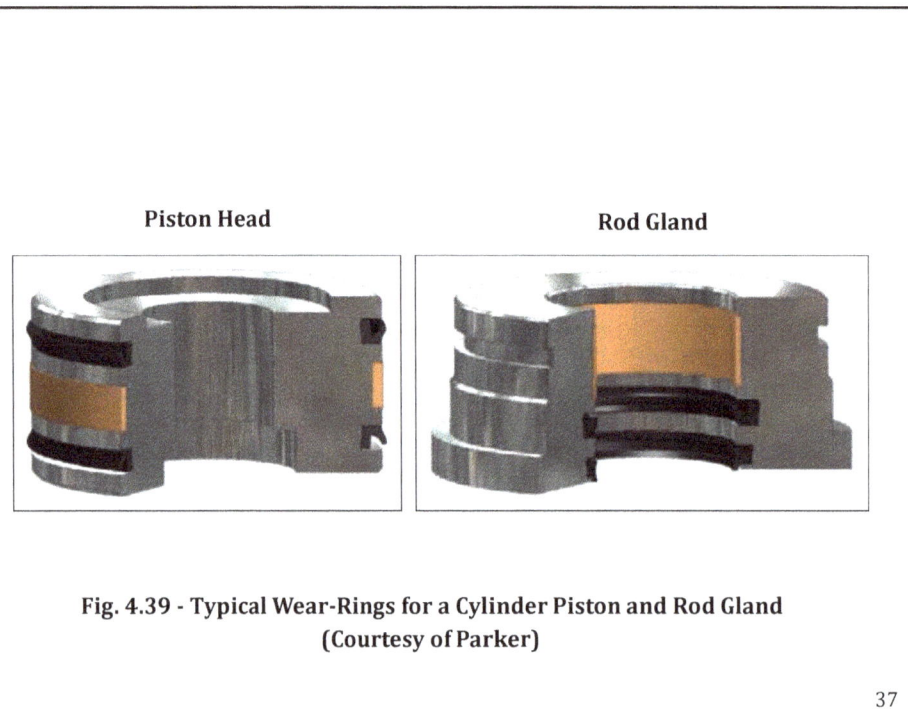

**Fig. 4.39 - Typical Wear-Rings for a Cylinder Piston and Rod Gland
(Courtesy of Parker)**

37

37

4.9- Backup Rings

- **Purpose:** Prevent seal extrusion.
- **Material:** Variety of materials (leather, rubber, elastomers and Teflon).
- **Size:** Wide range of sizes.

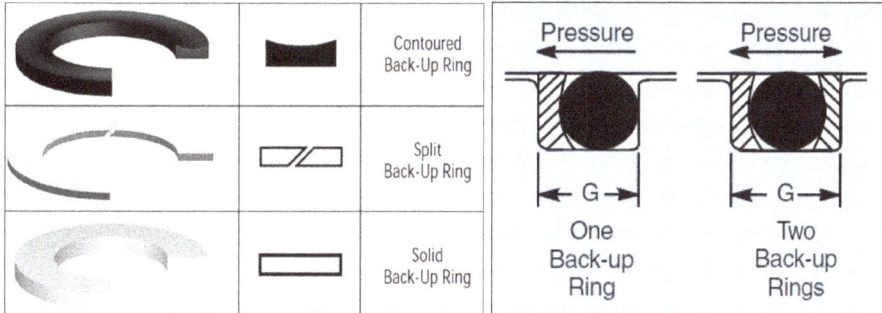

Fig. 4.40 – Cross Sections Standard Backup Rings (www.mfpseals.com)

Fig. 4.41 - Examples of Sealing Rings Supported by Backup Rings (Courtesy of Parker)

38

38

Piston Head

Rod Gland

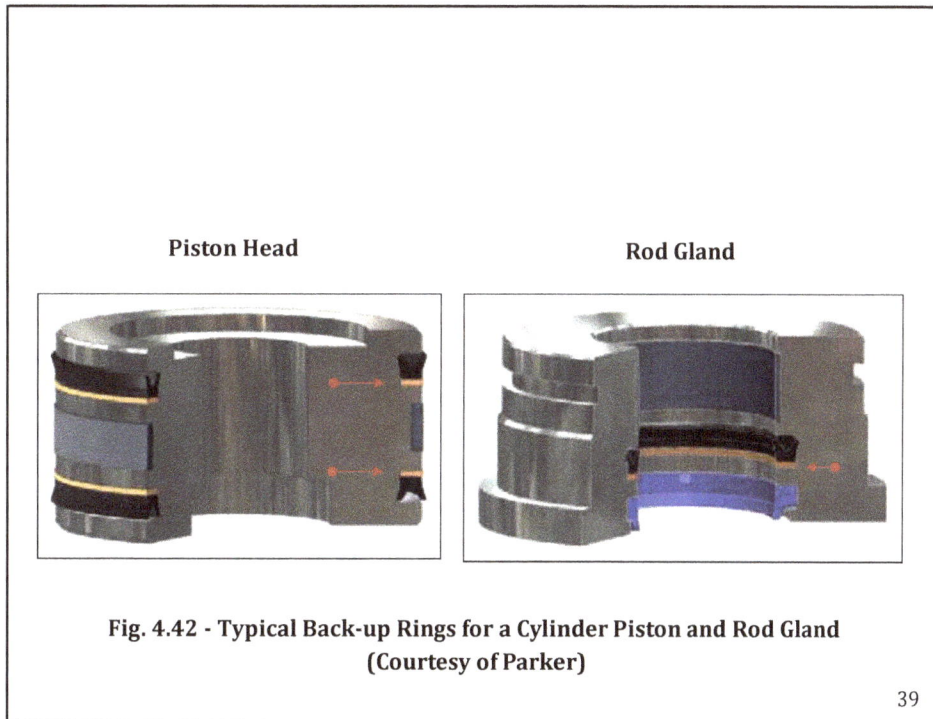

Fig. 4.42 - Typical Back-up Rings for a Cylinder Piston and Rod Gland (Courtesy of Parker)

39

39

4.10- Rod Wipers

- Contaminated hydraulic fluid → system failures.
- Rod Wipers (also known as Scrapers, Excluders or Dust Seals).
- **Purpose:**
  - o The first line of defense against contaminated cylinder rod.
  - o The last line of defense against external leaking.
  - o Host sufficient lubrication film for rod continuous self-lubrication.
- **Material:** (brass + other material)
- → compromised hardness and the flexibility.

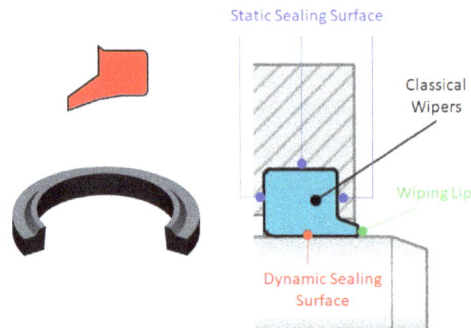

Fig. 4.43 - Traditional Single-Acting Rod Wiper Seal

40

40

Example 1: Single-Acting & Single-Lip Wiper Seal

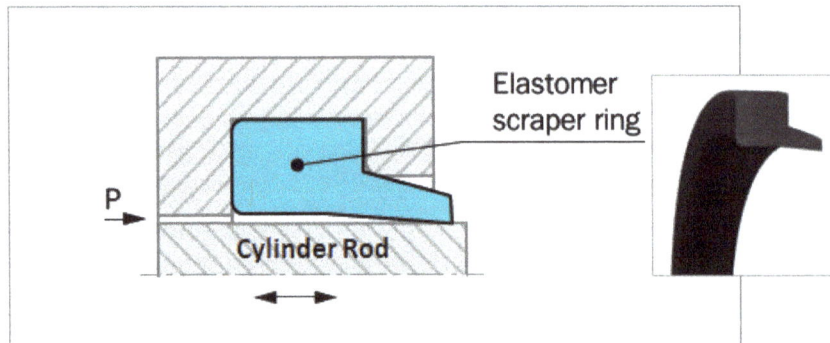

Fig. 4.44 - Typical Basic Single-Acting Rod Wiper Seal

(Courtesy of Trelleborg)

41

41

Example 2: Single-Acting and Redundant Lip Scraper.

- Modern Rod Wiper Seals → more than one scraper lip.

- Metallic lip → (self-adjusting + lasts longer + removes contaminants).

- Rubber lip → (last line of defense against contaminants + better sealing).

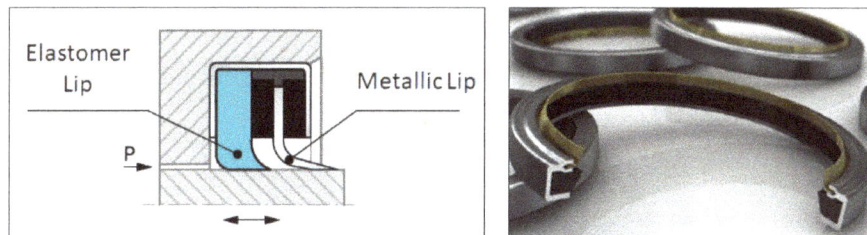

**Fig. 4.45 - Single-Acting Wipers with Redundant Sealing Lips
(Courtesy of Trelleborg)**

42

42

Example 3: Double-Acting Wiper Seals:
- Radial Squeeze with Two dynamic sealing lips.
- One lip for scrapping off dirt, particles, dust and water.
- One lip for sealing & lubrication.

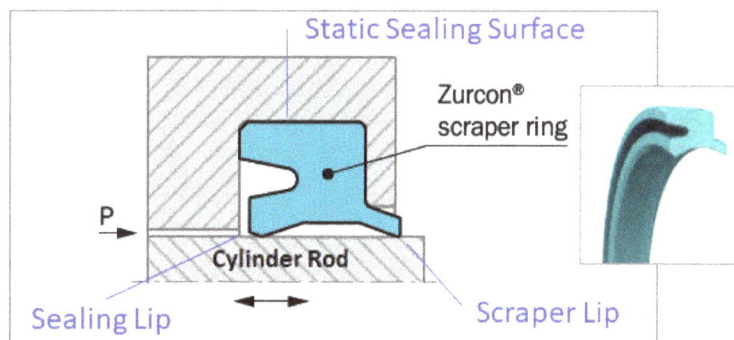

**Fig. 4.46 - Double-Acting with Redundant Sealing Lips
(Courtesy of Trelleborg)**

43

43

Example 4: High-Performance Rod Wiper
- **Known As** →"Umbrella Wiper Technology".
- **Applications** → harsh environment and numerous contaminants (agriculture, off-highway, and forestry equipment).
- **Design** → protective guard that entirely cover the retaining groove.
- **Advantages** →
  o Effectively remove water, debris, moisture, and mud.
  o Eliminate corrosion in wiper and gland housing.
  o Installed without special tools.
  o Made from material with good resistance to UV and chemical.
  o Offers long life and sealability.

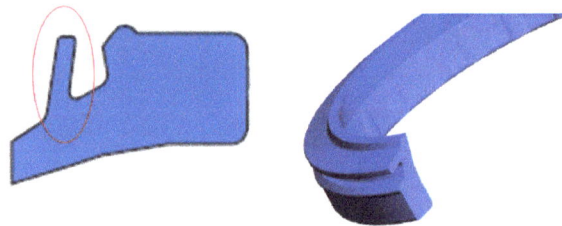

Fig. 4.47 - Umbrella Wiper Technology
(Courtesy of Hallite Seals)

44

44

4.11- Materials for Hydraulic Sealing Elements

- Seal Materials: Performance, service life, component's reliability.

- Conditions: T, P, fluids, environment, and side loads.

- Design: Modeling, simulation, field testing, trial & error.

- Codes and standards: ISO, Din, ASTM, and Manufacturer-based.

- Seals are produced from one or combination of:
  o Fabric.
  o Rubber (natural and synthetic).
  o Leather.
  o Metal.
  o Elastomeric Compounds (a mixture of base polymer and other chemicals that form a finished rubber material).
  o Engineered Plastics.

45

45

- A hydraulic seal composed of a (base compound & coating).
- PTFE is the most common coatings.
- Advantages of PTFE coating:
  - low frictional motion →
  - Ease of seal installation + eliminate sticking offer + reduce power

PolyteTraFluoroEthylene (PTFE)

Fig. 4.48 - PTFE Seal Material

46

46

ELASTOMER RUBBER COMPOUNDSAND REFERENCES					
General Description	**Chemical Description**	**Abbreviation (ASTM 1418)**	**ISO/DIN 1629**	**Other Trade Names & Abbreviations**	**ASTM D2000 Designation**
Nitrile	Acrylonitrile-butadiene rubber	NBR	NBR	Buna-N	BF, BG, BK, CH
Hydrogenated Nitrile	Hydrogenated Acrylonitrile-butadiene rubber	HNBR	(HNBR)	HNBR, HSN	DH
Ethylene-Propylene	Ethylene propylene diene rubber	EPDM	EPDM	EP, EPT, EPR	BA, CA, DA
Fluorocarbon	Fluorocarbon Rubber	FKM	FPM	Viton ®, Fluorel ®	HK
Chloroprene	Chloroprene rubber	CR	CR	Neoprene	BC, BE
Silicone	Silicone rubber	VMQ	VMQ	PVMQ FC, FE, GE	FC, FE, GE
Fluor-silicone	Fluor-silicone rubber	FVMQ	FVMQ	FVMQ	FK
Polyacrylate	Polyacrylate rubber	ACM	ACM	ACM	EH
Ethylene Acrylic	Ethylene Acrylic rubber	AEM	AEM	Vamac ®	EE, EF, EG, EA
Styrene-butadiene	Styrene-butadiene rubber	SBR	SBR	SBR	AA, BA
Polyurethane	Polyester urethane / Polyether urethane	AU / EU	AU / EU	AU / EU	BG
Natural rubber Natural rubber	Natural rubber Natural rubber	NR	NR	NR	AA

Vamac ® and Viton ® are registered trademarks of E. I. du Pont de Nemours and Company or affiliates.
Fluorel ® is a registered trademark of Dyneon LLC

Table 4.1 - Standard Abbreviations for Synthetic Rubber (news.ewmfg.com)

47

47

4.12- Properties and Test Methods for Hydraulic Sealing Elements

- **Scope:**
- Machines become faster & operating conditions become more severe.
  - → Seals operation becomes complex
  - → Seals material affects (static and/or dynamic) sealing function.
  - → Demand for seal material development and testing ↑.
  - → Consider combined disciplines (physics, mechanics, thermodynamics, fluid dynamics, tribology, etc.).

- **Standards:**
  - Some test methods are standardized.
  - Others are based on manufacturer R&D activities.

48

48

To obtain uniform and repeatable results →

- **Test Specimens**
  - Predefined standard specimens.
  - Two or more specimens are required.
  - Different specimens → rarely 100% agree.

- **Test Variables**
  Identical test variables → comparable results:
  - Temperature under which the test was performed.
  - Load or pressure used in the test.
  - Fluid medium in which the seal is in contact with during the test.
  - Duration and rate of applying the test procedure.
  - Environmental conditions, such as humidity in the air.

49

49

Hydraulic Sealing Elements and Test Methods
Change in Seal Shape
1- Resilience
Test Method: Non-Standard Test
Change in Seal Length
2- Modulus of Elasticity
3- Elongation
4- Yield Tensile Strength
5- Ultimate and Fracture Tensile Strengths
Test Method: Standard Test Method (ASTM D412 / DIN 53504)
Change in the Seal Cut
6- Tear Strength:
Test Method: Standard Test Method (ISO 34-1 / DIN 53507)
Change in Seal Thickness
7- Compression Set
Test Method: Standard Test Method (ASTM D395 / DIN ISO 815)
Change in Seal Volume
8- Swelling
Test Method: Standard Test Method (DIN ISO 1817)
9- Shrinkage
Test Method: Standard Test Method (TR-10 Low Temperature ASTM D1329-16)

Table 4.2 - Hydraulic Sealing Elements Properties and Test Methods 50

50

Change in Seal Surface
10- Surface Hardness
Test Method: Standard Test Method (ASTM D 2240 / ISO 868 / DIN 53505)
Change in Seal Chemical Structure
11- Compatibility with Hydraulic Fluid
Test Method: Standard Test Methods (ASTM D6546-15 OR ISO 6072)
Change in Seal Performance
12-Exrusion Resistance
Test Method: Standard Test Method (ASTM C1183 / C1183M)
13-Expolosive Decompression Resistance
Test Method: Standard Test Method (NACE TMO192-98)
14- Seal Friction
Test Method: Non-Standard Test
15- Wiper Performance
Test Method: Non-Standard Test

Table 4.2 - Continue

51

51

4.12.1- Resilience

❖ **Definition:**

Ability to return to original **shape** after a temporary deflection.

❖ **Units:** It is dimensionless property

❖ **Effect:**
- Resilience is primarily an inherent property of the elastomer.
- Reasonable resilience is vital for moving seals.

❖ **Testing:**
- It is very difficult to create a standard test method for this property →
- Functional testing under actual service conditions are considered.
- Visual inspection after test.
- It can be improved by compounding.

52

52

4.12.2- Modulus of Elasticity

❖ **Definition:**

Also called Young's modulus **E** = Stress/Strain

= stress acting on the seal body/corresponding % elongation).

❖ **Units:**
- Metric: MPa/(predetermined % elongation)
- English: psi/predetermined % elongation.

❖ **Effect and Importance:**
- Modulus of Elasticity of a compound ↑ →
  - Resilience ↑
  - Resistance to extrusion ↑

- Modulus of elasticity α seal hardness.

- Polyurethane and filled PTFE compounds →
  - High modulus of elasticity.

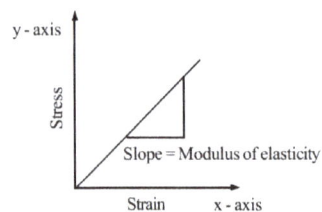

Fig. 4.49 – Modulus of Elasticity

53

53

4.12.3- Elongation

❖ **Definitions:**

% (increase in **length**/original length).

❖ **Units:**

▪ %.

❖ **Effect and Importance:**

▪ Ability of a seal to stretch → easy installation BUT easy to extrude.
▪ Change in the length overtime → is a clear sign of material degradation.

54

54

4.12.4- Yield Tensile Strength

❖ **Definitions:**

The stress within which the seal maintain elastic performance.

❖ **Units:**

▪ Metric (MPa).
▪ English (psi).

❖ **Effect and Importance:**

▪ This property determines the maximum strength beyond which the seal is plastically deformed → maximum load the seal can withstand.

55

55

4.12.5- Ultimate and Fracture Tensile Strength

❖ **Definitions:**
- Ultimate Tensile Strength → ultimate <u>elongation</u> + seal <u>nicking</u> starts.
- Fracture Tensile Strength → seal <u>breaks</u>.

❖ **Units:**
- Metric (Mega Pascals).
- English (psi).

❖ **Effect and Importance:**
- All properties (Elongation, Tensile Strength, Ultimate Strength, and Fracture Strength) →
  - Determine the application load conditions for sealing elements.
  - Used as quality assurance measures on production batches of elastomer materials.

56

56

❖ **Testing: (Standard Test Method ASTM D412 / DIN 53504)**

📹 Video 326 (1 min)

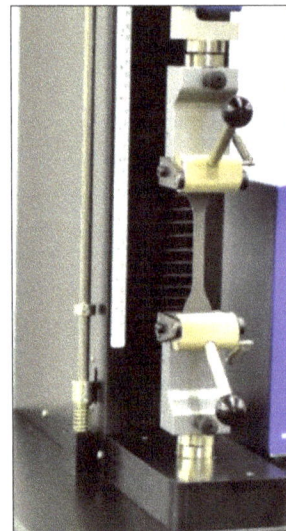

- <u>Test Purpose:</u> Determination of Tensile Strength values.

- <u>Test Specimen:</u>
  - Standardized tensile bars or rings with specific rectangular cross-sections.
  - Specimens with various cross sections → Inconsistent results.

- <u>Test Sequence:</u>
  - Test specimen is stretched at a constant speed to break.
  - ASTM D412 → Uniform pulling rate of 508 mm (20 inches) per minute.
  - Develop Tensile Strength curve.
  - Humidity in the air ↑→ tensile strength ↓

Fig. 4.50 - Elongation and Tensile Strength Test Machine 57

57

- Test Evaluation:
o Yield Tensile Strength, Ultimate Tensile Strength, Fracture Tensile Strength, and Elongation at break.

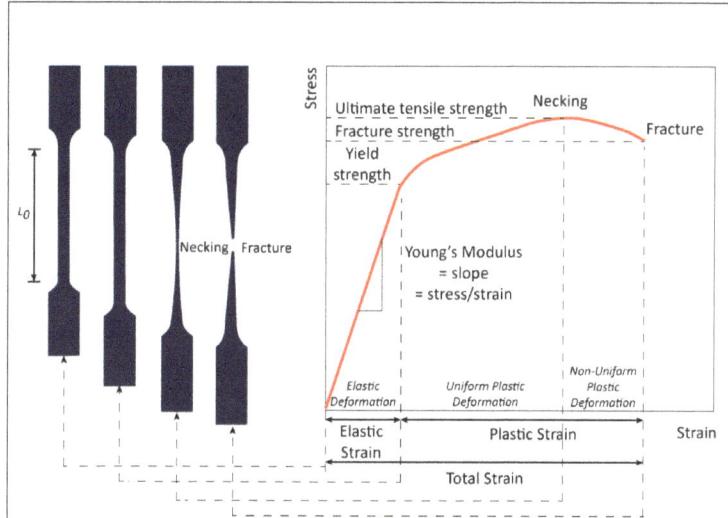

Fig. 4.51 - Results from Elongation and Tensile Strength Test

58

58

4.12.6 -Tear Strength

❖ **Definitions:**

Ratio
(force achieved at the moment of rupture/initial cross-section thickness).

❖ **Units:**
- Metric: kN/m.
- English: lb/in.

❖ **Effect and Importance:**
- Determines the ability to resist tear during assembly and operation.

- Once a crack is started → Seal fail quickly.

- Poor tear resistance (less than 100 lbs./in.) (17.5 kN/m) →:

- Danger of tearing during assembly if the seal pass over sharp edges.

59

59

❖ **Testing (Standard Test Method ISO 34-1 / DIN 53507):**

Video 327 (1 min)

▪ Test Purpose: Determination tear resistance.

▪ Test Specimen: Same as used in tensile strength.

▪ Test Sequence:
o A longitudinal cut is made in the material to be tested.
o The two half-strips are clamped in a pulling machine and pulled apart.

▪ Test Evaluation:
o Measure (force required to propagate the cut / sample thickness).

Fig. 4.52 - Tear Resistance Test

60

60

4.12.7- Compression Set

❖ **Definition:**
The ability to restore **thickness** and consequently the sealing force after a certain time in contact with a fluid medium under certain temperature.

❖ **Units:** % of a targeted value of deformation.

❖ **Effect and Importance:**
▪ Magnitude of CS ↓ → better restoring, sealing, and seal lifetime ↑.
▪ DO NOT get Confused: Poor CS → better recovery.
▪ Compression Set of a seal is affected by:
o Seal material.
o Improper gland design.
o Overtightening → over squeezing.
o Fluid incompatibility → swelling due to fluid.
o Temperature ↑→ seal hardness ↑→ elasticity ↓→ CS ↑ and restoring ↓.
o Incomplete curing (vulcanization) of seal material during production.

61

- **Example 1:** Poor CS → CS ≈ 100% → no recovery → flattened seal.

Fig. 4.53 - Flattened O-Ring due to Poor Compression Set (Courtesy of Parker)

- **Example 2:** Perfect CS → CS ≈ 0% → full recovery.

Fig. 4.54 - Seal Exhibiting nearly 100% Compression Set (Courtesy of Parker) 62

62

Testing (Standard Test Method ASTM D395 / DIN ISO 815):

- Test Purpose: Determination of the compression set.

- Test Specimen:
  - A specimen with a specific form & shouldn't be stressed previously.

- Test Sequence:
  - Must not be done earlier than 16 hours after elastomer manufacturing.
  - Measure original thickness (h_0).
  - The sample is compressed to the thickness (h_1) (default: $h_1 = 0.75\ h_0$).
  - The compressed specimen is stored
    - ✓ Medium (default Air).
    - ✓ Predefined T (should be mentioned).
    - ✓ Specified time (default: 24 or 72 hours).
  - Measure recovered thickness of the specimen (h_s) after 30 minutes of removal from storage and releasing the compression.

63

- Test Evaluation:

Video 328 (1 min)

- CS = 0% → Full recovery.
- CS = 100% → No recovery.

$$CS\ (\%) = \frac{(h_O - h_S)}{(h_O - h_1)} \times 100 \qquad 4.2$$

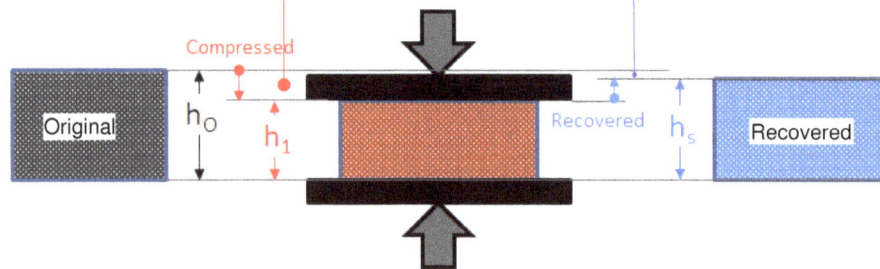

Fig. 4.55 - Compression Set Test

64

4.12.8- Swelling

❖ **Definition:**
- Coefficient of thermal expansion of elastomers > steel →
- Elastomer expand more than the gland when it gets hot → Swelling.
- Swelling: **Volumetric increase** when the seal get in contact with fluid.

❖ **Units:** It is expressed as a % of the original volume.

❖ **Effect and Importance:**
- Possibility of seal extrusion under high pressure.
- Reduced hardness, elasticity, and tensile strength.

❖ **Seal Swelling depends on:**
1. Fluid in contact.
2. Composition of the elastomeric compound.
3. Working conditions (temperature, pressure, humidity, and time).
4. Geometric form (thickness) of the seal.
5. Stress condition of the seal.

65

❖ **Testing (Standard Test Method DIN ISO 1817):**
▪ Test Purpose: Determination of seal volume increase after a defined storage period in contact with specific fluid under certain temperature.

▪ Test Sequence:
o Measure the original volume of the seal.
o The seal is immersed in a test fluid and stored according to the standard or to customer specifications.
o At the end of the storage period (and after cooling down), the volume of the seal is measured again.
o The result is expressed as a percentage of the initial volume.

▪ Test Evaluation: As a rule-of-thumb (unless otherwise stated).
o Static seals → up to 50% swell is tolerated.
o Dynamic seals → (10-20) % swell is a reasonable.
o Seals with smaller cross-sections → swell more than larger ones.

66

66

4.12.9- Shrinkage

❖ **Definition:**
▪ Coefficient of thermal expansion of elastomers > steel →
▪ Elastomer shrink more than the gland when it gets cold → Shrinkage.
▪ Shrinkage: **Volumetric Decrease** when the seal get in contact with fluid.

❖ **Units:** It is expressed as a % of the original volume.

❖ **Effect and Importance:**
▪ Shrinkage → Reduced retaining force between the seal and the static housing faces.
▪ Shrinkage → possible leakage.

67

67

❖ **Testing (Low-Temp. Standard Test Method) ASTM D1329-16/TR-10):**

- Test Purpose: Determination of elastomer retraction and viscoelastic properties at low temperature.

- Test Specimen:
  o Standard specimens have lengths of 1", 1.5", or 2".

- Test Sequence:
  o A small dog-boned specimen is held in an elongated condition.
  o The test fixture is used to apply 100% elongation.
  o The specimen is subjected to a low temperature for certain specified time.
  o The specimen is then allowed to retract freely while raising the temperature at a uniform rate of 1°C (1.8 °F) per minute.
  o Measure the temperature at 10% and 70% retraction (shrinkage).
  o Comparison 1: % shrinkage at a specific T.
  o Comparison 2: T at specific % shrinkage.

68

68

- Test Evaluation:
  o Elastomers are compared based on temperatures corresponding to 10% and 75% shrinkage.
  o Maximum tolerated volumetric shrinkage for both static and dynamic seals is (3-4) %.

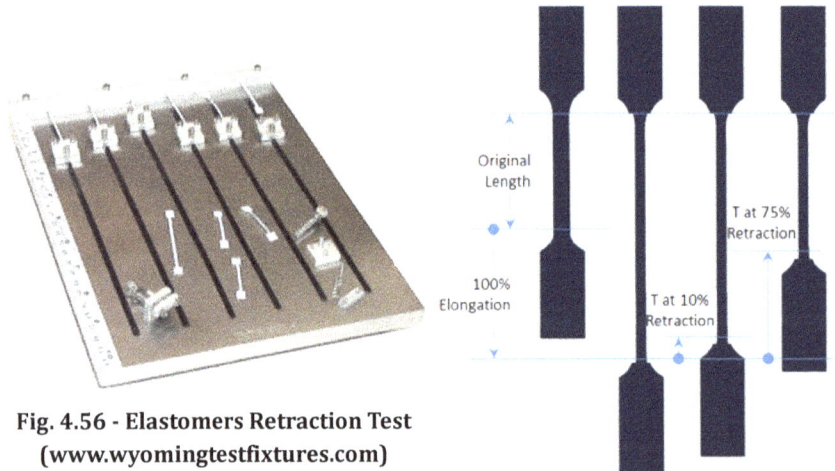

**Fig. 4.56 - Elastomers Retraction Test
(www.wyomingtestfixtures.com)**

69

69

4.12.10- Surface Hardness

❖ **Definition:**
- It is the resistance of a body against penetration of a harder body of a standard shape at a defined load.

❖ **Units:**
- The hardness scale has a range of 0 (softest) to 100 (hardest).
- Shore A (for soft-to-medium compounds) and
- Shore D for (for medium-to-hard compounds).

❖ **Effect and Importance:**
- Hardness in Dynamic Seals ↑→ resistance to abrasion, wear, and scraping ↑.
- Hardness in Static Seals ↑→ seal extrusion ↓.
- Seal hardness ↑→ Modulus of Elasticity ↑ and Leakage ↑.
- Seal hardness ↑→ Seal friction ↓.

70

70

❖ **Testing:**
- Standards:
  - For Shore A" and "Shore D" →
  - **Standard Test Methods ASTM D 2240 / ISO 868 / DIN 53505.**

- Test Purpose: determination of seal hardness

- Test Specimen (Rubber Desk):
  - Diameter min. 30 mm (1.181 inch).
  - Thickness min. 6 mm (0.240 inch).
  - Surfaces (upper and lower) should be smooth and flat.
  - The test samples must not be previously stressed.

71

71

- Hardness Testers:
  - Manual or automated Hardness Tester, called "Durometer".
  - A calibrated spring force is applied on a standard pin (Indentor) against the specimen.
  - 1-degree Shore (A) → Pin Penetration = 0.001-inch (0.0254 mm).
  - Harder material → penetration resistance ↑→ pin deflection and spring compression ↑→ shore number↑.

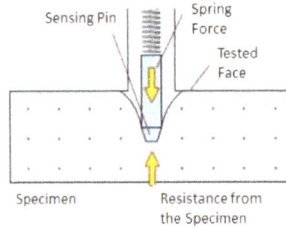

Video 329 (6 min)

Sensing Pin

Spring Force

Tested Face

Specimen

Resistance from the Specimen

Fig. 4.57 - Hardness Test Devices for Elastomers

72

72

- Shore A (soft-to-medium compounds) → conical head.
- Shore D (medium-to-hard compounds) → pin with spike head.

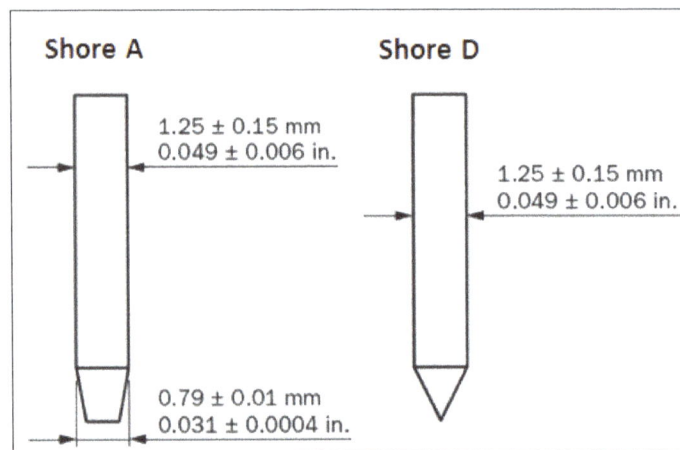

Shore A

1.25 ± 0.15 mm
0.049 ± 0.006 in.

0.79 ± 0.01 mm
0.031 ± 0.0004 in.

Shore D

1.25 ± 0.15 mm
0.049 ± 0.006 in.

Fig. 4.58 - Pins for Hardness Tests (Courtesy of Trelleborg)

73

73

- Test Sequence:
  - Must not be carried out earlier than 16-Hrs after manufacturing.
  - Test Temperatures =23 ±2 °C (73.4 ±2 °F).
  - Test Temperatures should be mentioned in the test report.
  - Apply the spring load on the pin at a certain speed.
  - The value is read after a holding time of three seconds.
  - **Note:** O-Ring → no flat surface → measuring spot at the actual crown of the O-Ring, the point that gives the most reliable reading.

Fig. 4.59 - Hardness Test for O-Rings

74

74

- Test Evaluation (as shown in Table 4.3):
  - Shore A = 60-75 → recommended for O-rings.
  - Shore A = 75-90 → recommended for seals of high resistant to abrasion.
  - Shore A > 90 → Shore D scale is used for harder compounds.
  - Shore A 90 ≈ Shore D 40.

Shore A															
5	10	20	30	40	50	60	70	80	90						
Shore D															
									40	50	60	70	80	85	
Very Soft			Soft			Medium				Hard					

Table 4.3 - Hardness Scales for Hydraulic Seals

75

75

4.12.11- Compatibility with Hydraulic Fluids

❖ **Definition:**

▪ The ability of a seal to resist chemical interaction with the working hydraulic fluids.

❖ **Effect and Importance:**

▪ Any change in the seals chemical structure →

▪ Affects the seal shape and physical properties →

▪ Seal Performance.

▪ Seal deterioration due to chemical reaction →

▪ Clogging of control orifices, leakage, and the relevant consequences.

76

Seal materials	Fluid Types					
	Petroleum oil	Water-in-Oil Emulsion	Water Glycol	Phosphate Ester*	Chlorinated hydrocarbon	Synthetic with petroleum fractions
Buna-N (Acrylonitrile)	Excellent	Excellent	Very Good	Poor	Poor	Poor
Neoprene (Chloroprene)	Good	Good	Good	Poor	Poor	Poor
Butyl	Poor	Poor	Good	Fair to good	Poor	Poor
Silicone	Fair	Fair	Fair to poor	Fair to good	Poor to fair	Fair
Ethelene-Propylene	Poor	Poor	Good to excellent	Excellent	Fair	Poor
Viton® (Fluorocarbon)	Excellent	Excellent	Excellent	Good to Excellent	Good to Excellent	Good to Excellent
Metals	Conventional	Conventional	**	Conventional	Conventional	Conventional
Pipe Sealants	Conventional, Loctite® or Teflon® tape	Conventional, Loctite® or Teflon® tape	Loctite® or Teflon® tape	Loctite® or Teflon® tape	Loctite® or Teflon® tape	Loctite® or Teflon® tape

▪ *Many types and blends of fluids are sold under the designation "phosphate ester." Check with fluid supplier to verify exact compatibility.
▪ **Avoid zinc, cadmium, or galvanized materials.
▪ Viton® and Teflon® are trademarks of E.I DuPont DeNemours & Co., Inc.
▪ Loctite® is a trademark of the Loctite Corp.

Table 4.4 - Compatibility of Common Hydraulic Fluids with Common Seal Materials (www.schoolcraftpublishing.com)

77

❖ **Testing (Standard Test Methods ASTM D6546-15 and ISO 6072):**

▪ Test Purpose: Determining compatibility of elastomeric seals for industrial hydraulic fluid applications.

▪ Test Procedure: Exposing a specimen to industrial hydraulic fluids under definite conditions of temperature and time.

▪ Test Evaluation: By comparison to a new specimen (work function, hardness, physical properties, compression set, and seal volume).

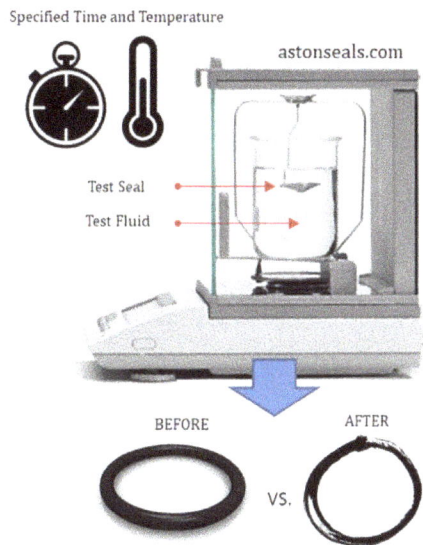

Fig. 4.60 - Fluid Compatibility Standard Test Method   78

78

4.12.12- Extrusion Resistance

❖ **Definition:**
▪ The seal ability to resist extrusion through the gap between sealed surfaces.

❖ **Effect and Importance:**
▪ Working pressure or seal material softness → Seal extrusion →
▪ Seal failure, excessive leakage, and oil contamination.

❖ **Testing: (Standard Test Method ASTM C1183):**
▪ Test Purpose: determining the extrusion rate of elastomeric sealants.

▪ Test Procedure: This test method measures the volume of sealant extruded over a given period of time at a given pressure (kPa or psi).

79

79

4.12.13- Explosive Decompression Resistance

❖ **Definition:**
- The seal ability to resist damage due to releasing of absorbed gas when pressure is suddenly reduced.

❖ **Effect and Importance:**
- Explosive decompression → seal damage and improper functioning.

❖ **Testing: (Standard Test Method NACE TMO192-98):**
- This test is used for evaluating elastomeric materials in Carbon Dioxide decompression environments.

80

80

4.12.14- Hydraulic Seal Friction
4.12.14.1- Hydraulic Seal Friction Conditions

- Seal Friction Modes based on the lubrication condition between the seal and the sealed surface:

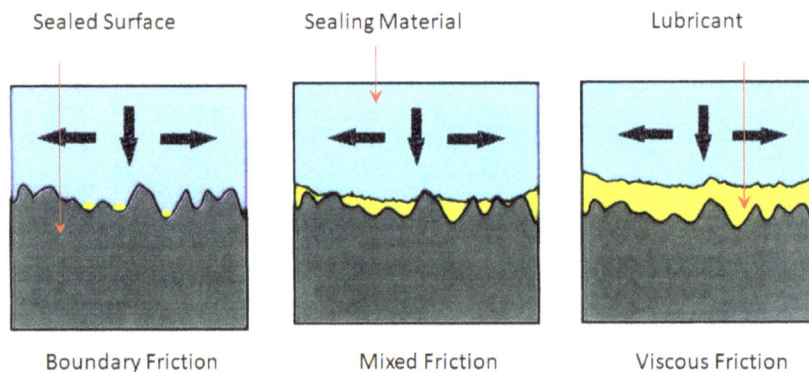

Fig. 4.61 - Hydraulic Seal Friction Conditions

81

81

4.12.14.2- Friction in Translational Seals

- Example of translational friction → cylinder rod and piston seals.

- This friction is mistakenly ignored in calculating a cylinder F or P.

- Seal friction in cylinders:

    at a speed 5-15 cm/s (10 – 30 fpm) ≈ 5% of the cylinder input power.

- Seal friction → responsible for 1/10 of cylinder leakage.

82

82

- <u>Experimental Investigation:</u>
- Cylinder extended at 75 bar (1088 psi) constant pressure.
- Seal friction drops once the rod starts extending, then gradually increases with increasing velocity.

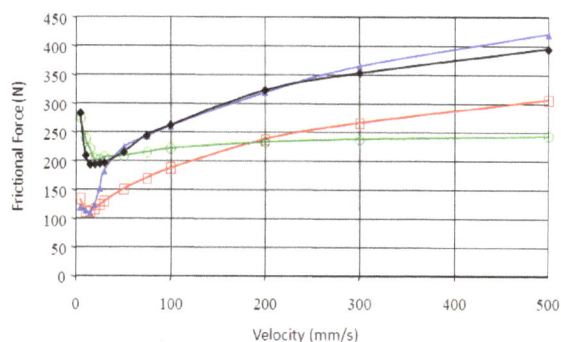

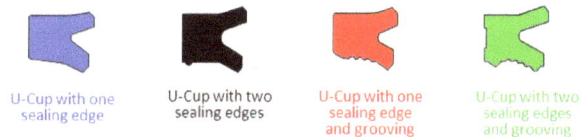

Fig. 4.62 - Rod Seal Friction at Constant Pressure   83

83

4.12.14.3- Friction in Rotational Seals

- Examples of rotational seals → pumps and motors shaft seals.

$$v = \omega \times R = 2\pi n \times R \qquad\qquad 4.3$$

<u>Where:</u>
v = Tangential speed in (m/min) or (FPM) depends on the units of **R**.
R = Radius of the shaft in (m) or (foot).
n = Shaft rotational speed (RPM).

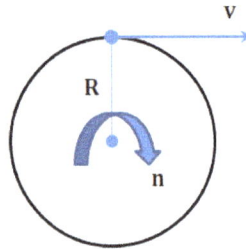

Fig. 4.63- Surface (Tangential) Speed of a Rotational Shaft

84

84

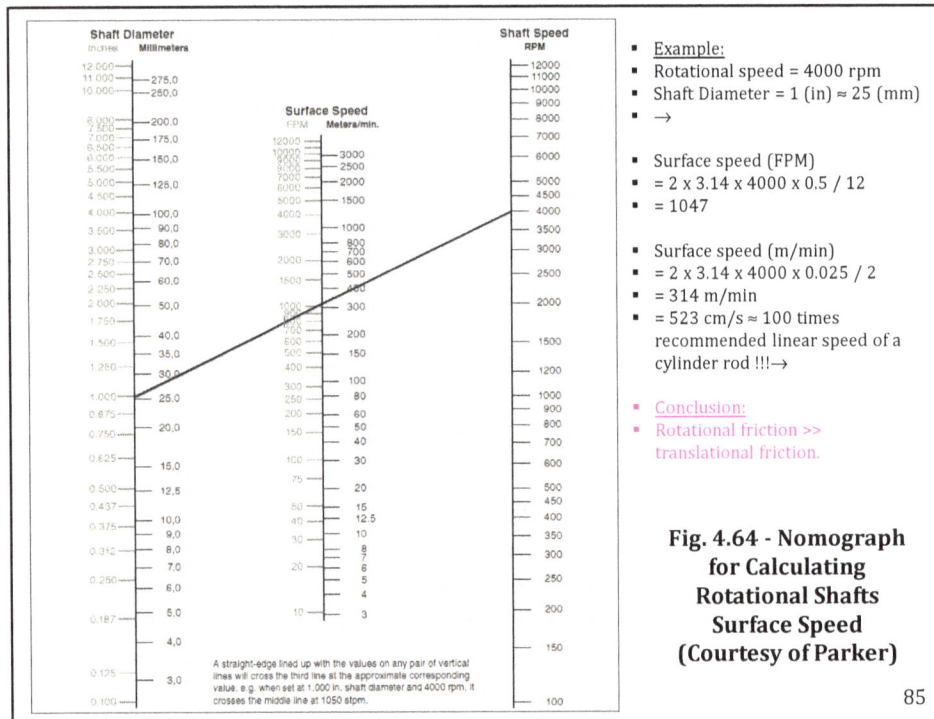

- <u>Example:</u>
- Rotational speed = 4000 rpm
- Shaft Diameter = 1 (in) ≈ 25 (mm)
- →

- Surface speed (FPM)
- = 2 x 3.14 x 4000 x 0.5 / 12
- = 1047

- Surface speed (m/min)
- = 2 x 3.14 x 4000 x 0.025 / 2
- = 314 m/min
- = 523 cm/s ≈ 100 times recommended linear speed of a cylinder rod !!!→

- <u>Conclusion:</u>
- Rotational friction >> translational friction.

Fig. 4.64 - Nomograph for Calculating Rotational Shafts Surface Speed (Courtesy of Parker)

85

85

4.12.14.4- Factors Affecting Seal Friction

Friction of dynamic seals depends primarily on:

- **Seal Geometry:** shape, dimensions, exposed area, and dynamic surface roughness.

- **Seal Material:** type and properties.

- **Working Conditions:** temperature and pressure.

- **Speed (translational or rotational):**
  - **Low Speed:** → affects the performance of the seal.

  - **High Speed:** → breakdown of oil film → seal runs dry → premature seal failure.

86

86

4.12.14.5- Controlling Seal Friction

- ❖ Basically, seal friction can't be 100% eliminated.

- ❖ However, lowering seal friction →
- Energy loss ↓
- Heat generation ↓
- Seal wear ↓
- Rate of chemical attack ↓
- Seal extrusion resistance ↑
- Seal life ↑

87

87

❖ Seal friction can be reduced by:

- Seal Design:

  o Waved seal surfaces to retain lubricating oil.

  o Seal design and placement should consider better heat dissipation.

- Seal Material:

  o Coating seals by a dry lubricant such as Teflon layer.

  o Compromise (Seals hardness ↑→ fiction ↓ and seal leakage ↑).

- Working Conditions:

  o Select proper seals based on field working conditions.

  o Control working conditions within recommended limits.

  o Use compatible fluids with anti-friction additive packages.

88

4.12.15- Wiper Performance Test

❖ **Effect and Importance:**
- Cylinder rod seals → source of dirt ingression
- Dirt ingression → hydraulic system inefficiency, degradation and failure.

❖ **Testing:**
- Test Purpose:
  o Innovative non-standard test method developed by Hallite & MSOE →
  o Assess the amount of dirt entering through the rod wiper.

- Test Conditions:
  o Test Duration: 24.000 cycles, 24384 meters (80,000 feet) linear travel.
  o Cycle Rate: 0.25 Hz
  o Total Stroke Length: 101.6 (40 inches).
  o Test Pressure: Atmospheric
  o Test Temperature: 66 $^{\circ}$C (150 $^{\circ}$F).
  o Test Oil: MIL-PRF-46170
  o Test Contaminant: ISO 12103-1-A4 Course Test Dust

89

- Test Procedure:
1. A rod wiper is installed to simulate conditions in standard cylinder application.

2. Fluid is heated to test T.

3. Fluid is pumped beneath the rod wiper.

4. Air is supplied to dust chamber.

5. The rod is cycled with the specified frequency to a specified duration.

6. Oil drains back to reservoir.

7. Dirt content is measured by particle counter.

Video 443 (1 min)

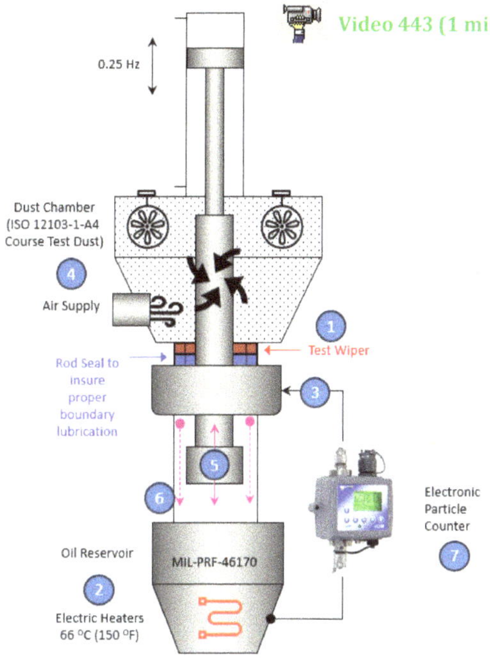

0.25 Hz

Dust Chamber (ISO 12103-1-A4 Course Test Dust)

Air Supply

Rod Seal to insure proper boundary lubrication

Test Wiper

Electronic Particle Counter

Oil Reservoir

MIL-PRF-46170

Electric Heaters 66 °C (150 °F)

Fig. 4.65 - Cylinder Rod Wiper Performance (Courtesy of Hallite Seals) 90

90

- Test Evaluation:
  o MSOE tested the Hallite wipers against two competitors.

  o Results→
  o Hallite → most protection & least amount of dirt ingression.

91

4.13- Best Practices for Hydraulic Seals Selection

❖ **Ideal** **hydraulic seal is the one that:**

- Performs efficiently with low friction.
- Works under extended operating conditions (pressure and temperature).
- Has high tensile strength and resists twisting and spiral failures.
- Has high tear and abrasion resistance.
- Compatible with various types of hydraulic fluids.
- Resists chemicals and acids.
- Requires minimum space and is easy to install.
- Cost effective.

92

92

❖ Unfortunately, there is no one seal that satisfies all the sealing requirements.
→ seal properties should be application-based compromised .

❖ **Best Practices for selecting hydraulic seals:**

1. Selection of Seal Type Based on Application.
2. Selection of Seal Dimensions.
3. Selection of Seal Lip Geometry.
4. Selection of Seal Crossectional Shape.
5. Selection of Seal Material Based on Working Temperature.
6. Selection of Seal Material Based on Working Pressure.
7. Selection of Seal Material Based on Working Fluid.
8. Selection of Seal Material Based on Hardness.
9. Selection of Seal Material Based on General Properties.

93

93

4.13.1- Selection of Seal Type

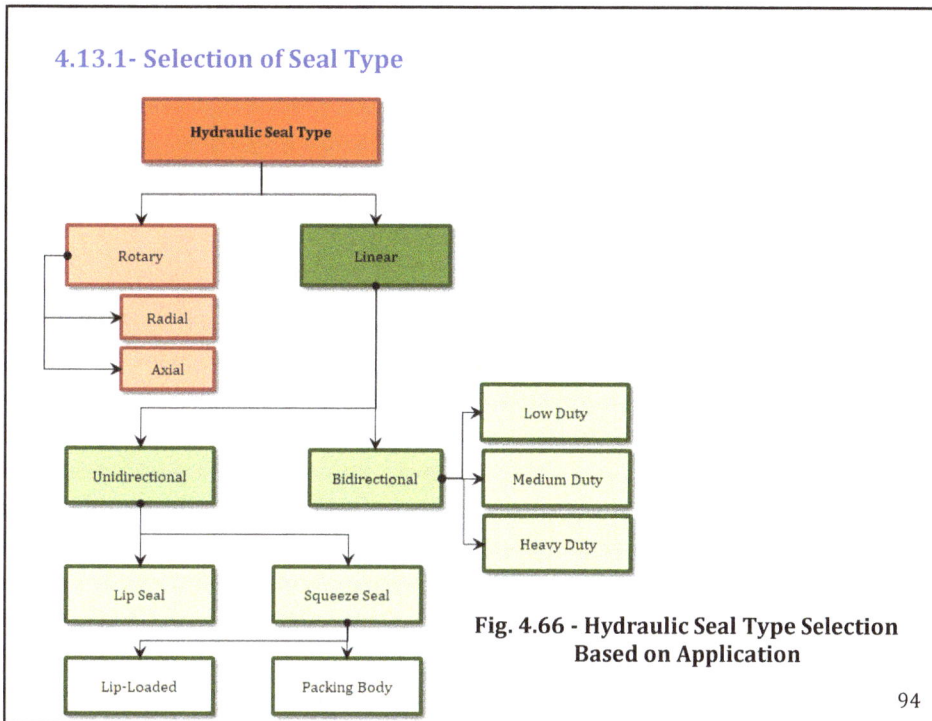

Fig. 4.66 - Hydraulic Seal Type Selection
Based on Application

94

94

4.13.2- Selection of Seal Dimensions

❖ **O-Ring Cross Section versus its Stability:**

▪ Several standard cross section diameters available →

▪ Smaller cross section →

   o are inherently less stable than larger cross sections, →

   o Twist in the groove when reciprocating motion occurs →

   o O-ring spiral failure and leakage.

95

95

4.13.3- Selection of Seal Lip Geometry

- Lip geometry → several functions:
- Sealing.
- lubricating film.
- Hydroplaning.
- contamination exclusion.

- **Example:**
- *Straight Cut*
- → Very high contamination exclusion.
- → Rod Wipers.

Table 4.5 - Effect of Lip Geometry on Seal Function (Courtesy of Parker)

Contact Shape	Rounded	Straight Cut	Beveled	Square
Seal Lip Shape / Shape of Contact Force/ Stress Profile				
Film Breaking Ability	Low	High	Very High	Medium
Contamination Exclusion	Low	Very High	Low	High
Tendency to Hydroplane	High	Very Low	Low	Medium
Typical Uses	Pneumatic U-cups	Wipers and Piston Seals	Rod Seals	Piston Seals

96

96

4.13.4- Selection of Seal Crossectional Shape

- Seal crossectional shape → affects sealability & friction.
- Seals are categorized as either "Lip Seals" or "Squeeze Seals".
- Other seals fall somewhere in between.
- Lip Seals → low friction, low wear, poor sealability at low P.
- Squeeze Seals → opposite
- In dynamic seals → sealability is traded off with low friction performance.

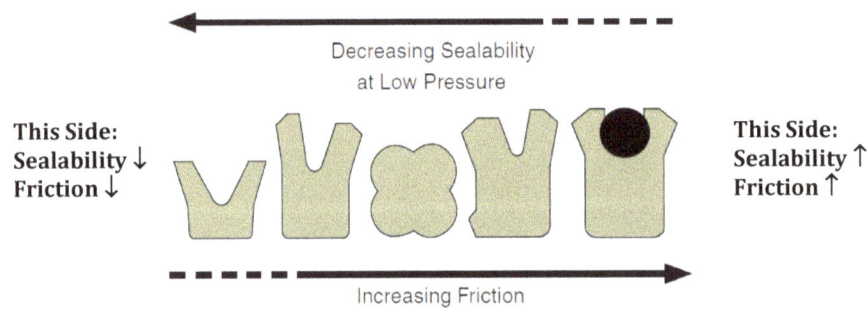

Decreasing Sealability at Low Pressure

This Side: Sealability ↓ Friction ↓

This Side: Sealability ↑ Friction ↑

Increasing Friction

Fig. 4.67 - Lip vs. Squeeze Sealing (Courtesy of Parker)

97

4.13.5- Selection of Seal Material Based on Working Pressure

- Consult seal manufacturer based on working pressure.

- Working pressure → affects the seal lifetime.

- Example:
  - A seal that is designed to work at 150 bar (2175 psi) can last for 2 years.
  - If it works at 200 bar (2900 psi) → it will last for 2 months.
  - If it works at 350 bar (5076) → it will last for 2 days.

98

98

4.13.6-Selection of Seal Material Based on Working Temperature
- Working temperature → affects the seal performance.
- **High Temperatures →**
- Oil at the interface surface evaporates → dry-running condition.
- Lack of lubrication → greatly accelerated seal wear → leakage ↑.
- Softening the seal → seal extrusion.
- Prolonged exposure to excessive heat → permanent hardening → leakage ↑.
- Thermal expansion → difficult to design the grooves for the seals.

- **Low Temperatures →**
- Loss of elasticity → damage during shrinkage.
- Seal hardness ↑ → glasslike brittleness → a dynamic seal may shatter if mechanically struck or hit with pressure spike.

- NBR common elastomeric materials → T = -50 to 100 °C (-58 to 212 °F).
- Silicon rubber (Viton Seals) → T -50 to 200 °C (-58 to 392 °F).
- Synthetic rubber →
  - Continuous use at high or low temperatures.
  - Tolerate temperature fluctuation. 99

99

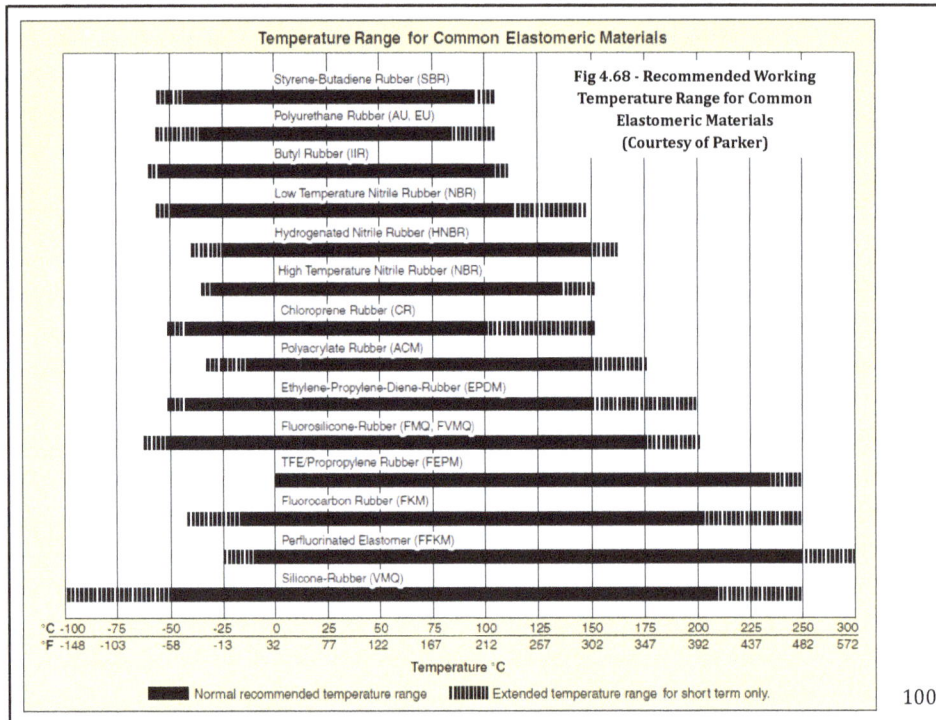

Fig 4.68 - Recommended Working Temperature Range for Common Elastomeric Materials (Courtesy of Parker)

100

4.13.7- Selection of Seal Material Based on Working Fluid

- Hydraulic sealing elements must not react chemically with the fluid.
- Water also can adversely affect the seals.
- Consult seal manufacturers.

Table 4.6 - Recommended Seal Materials for Fire-Resistant Fluids (Courtesy of Parker)

Properties of the Four Groups of Non-Flammable Fluids

Properties	HFA/HFB	HFC	HFD
kinematic viscosity (mm²/s) to 50°C (122°F)	0.3 to 2	20 to 70	12 to 50
viscosity/temperature relationship	good	very good	bad
density at 15°C (59°F)	ca. 0.99	1.04 to 1.09	1.15 to 1.45
temperature range	3°C to 55°C (37°F to 131°F)	-25°C to 60°C (-13°F to 140°F)	-20°C to 150°C (-4°F to 302°F)
water content (weight %)	80 to 98	35 to 55	none
stability	emulsion poor solution very good	very good	very good
life of bearings	5 to 10%	6 to 15%	50 to 100%
heat transfer	excellent	good	poor
lubrication	acceptable	good	excellent
corrosion resistance	poor to acceptable	good	excellent
combustion temperature	not possible	after vaporizing of water under 1000°C (1832°F)	ca. 600°C (1112°F)
environmental risk	emulsion: used oil synth.: dilution	special waste	special waste
regular inspection	pH-level concentration water hardness micro-organisms	viscosity water content pH-level	viscosity neutral pH spec. gravity
seal material	NBR, FKM	NBR	FKM, EPDM[1]

(1) only for pure (mineral oil free) phosphate-ester (HFD-R)

101

Seal Material	Compatible Fluids	Temperature Range
1. Metallic piston rings	Petroleum base and synthetic fluids, phosphate esters - for high pressure and severe conditions	Low to 500°F (260°C)
2. Leather	Petroleum base and some synthetics, phosphate esters - for medium to high pressure	-65°F to 225°F -54°C to 107°C
3. Neoprene rubber	General purpose industrial use, Freon™ 12; weather and salt water resistant	-65°F to 300°F -54°C to 149°C
4. Nitrile rubber (Buna N™)	Petroleum base fluids and mineral oils - used for some rotating seals, extrusion resistant	-65°F to 225°F -54°C to 107°C
5. Silicone rubber	Water and petroleum base fluids, phosphate esters; low tensile strength and tear resistance recommended for static seals only	-80°F to 450°F -62°C to 232°C
6. Fluoro-Elastomers (Viton™ and Fluorel™)	Petroleum base, synthetic, diester, silicate ester, and halogenated hydrocarbon fluids - for high temperature fluid applications	-20°F to 400°F -29°C to 204°C
7. Polyurethane	Petroleum base fluids - high resistance to ozone, sunlight and weathering; low water resistance	-65°F to 200°F -40°C to 93°C

**Table 4.7 - Fluid Compatibility for Common Sealing Materials
(Hydraulic Specialist Study Manual, IFPS)**

102

4.13.8- Selection of Seal Material Based on Hardness

- Seal hardness ↑ → seal friction ↓
- Seal hardness ↑ → leakage ↑.
- Seal hardness should be compromised based on the seal application.

Example 1:
- Guide-Rings → main objective is not to seal, but to guide a piston or a rod at low friction → high hardness is recommended.

Example 2:
- Piston Seals → low hardness is recommended → better sealability.

103

199/335

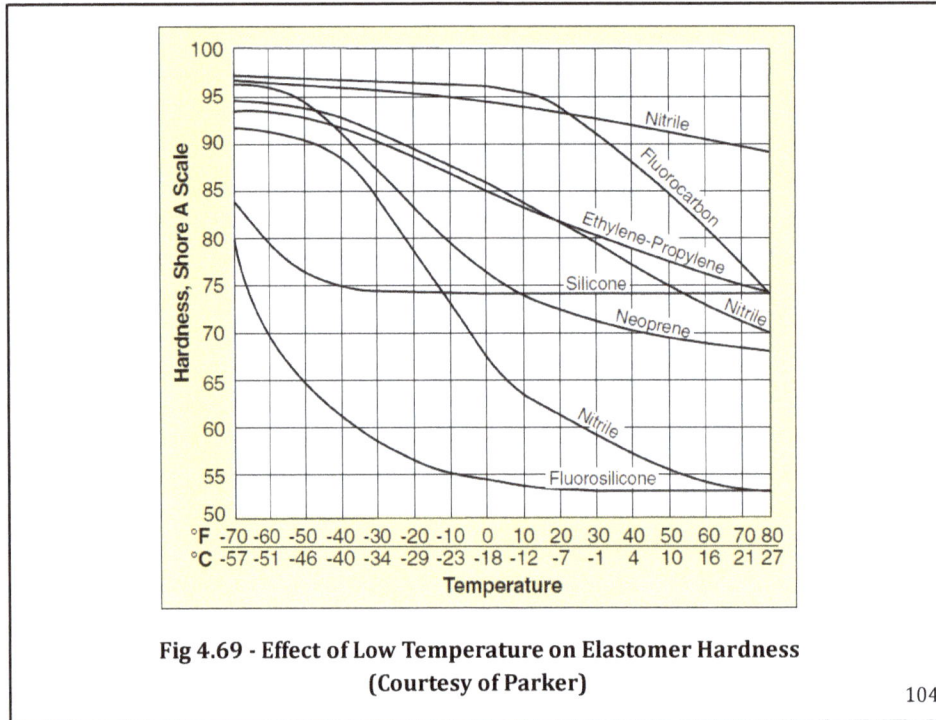

**Fig 4.69 - Effect of Low Temperature on Elastomer Hardness
(Courtesy of Parker)**

104

4.13.9- Selection of Seal Material Based on General Properties

Comparison of Properties of Commonly Used Elastomers
(P = Poor – F = Fair – G = Good – E = Excellent)

Elastomer Type (Polymer)	Parker Compound Prefix Letter	Abrasion Resistance	Acid Resistance	Chemical Resistance	Cold Resistance	Dynamic Properties	Electrical Properties	Flame Resistance	Heat Resistance	Impermeability	Oil Resistance	Ozone Resistance	Set Resistance	Tear Resistance	Tensile Strength	Water/Steam Resistance	Weather Resistance
AFLAS (TFE/Prop)	V	GE	E	E	P	G	E	E	E	G	E	E	PF	PF	FG	GE	E
Butadiene		E	FG	FG	G	F	G	P	F	F	P	P	G	GE	E	FG	F
Butyl	B	FG	G	E	G	F	G	P	G	E	P	GE	FG	G	G	G	GE
Chlorinated Polyethylene		G	F	FG	PF	G	G	GE	G	G	FG	E	F	FG	G	F	E
Chlorosulfonated Polyethylene		G	G	E	FG	F	F	G	G	G	F	E	F	G	F	F	E
Epichlorohydrin	Y	G	FG	G	GE	G	F	FG	FG	GE	E	E	PF	G	G	F	E
Ethylene Acrylic	A	F	F	FG	G	F	F	P	E	E	F	E	G	F	G	PF	E
Ethylene Propylene	E	GE	G	E	GE	GE	G	P	G	G	P	E	GE	GE	GE	E	E
Fluorocarbon	V	G	E	E	PF	GE	F	E	E	G	E	E	E	F	GE	F	E
Fluorosilicone	L	P	FG	E	GE	P	E	G	E	P	G	E	P	F	F	F	E
Isoprene		E	FG	FG	G	F	G	P	F	F	P	P	G	GE	E	FG	F
Natural Rubber		E	FG	FG	G	E	G	P	F	F	P	P	G	GE	E	FG	F
Neoprene	C	G	FG	FG	FG	F	F	G	G	G	FG	GE	F	FG	G	F	E
HNBR	N. K	G	E	FG	G	GE	F	P	E	G	E	G	GE	FG	E	G	G
Nitrile or Buna N	N	G	F	FG	G	GE	F	P	G	G	E	P	GE	GE	GE	FG	F
Perfluorinated Fluoroelastomer	V. F	P	E	E	PF	F	E	E	E	G	E	E	G	PF	FG	GE	E
Polyacrylate	A	G	P	P	P	F	F	P	E	E	E	E	F	FG	F	P	E
Polysulfide		P	P	G	G	F	F	P	P	E	E	E	P	P	F	P	E
Polyurethane	P	E	P	FG	G	E	FG	P	F	G	G	E	F	GE	E	P	E
SBR or Buna S		G	F	FG	G	G	G	P	FG	F	P	P	G	FG	GE	FG	F
Silicone	S	P	FG	GE	E	P	E	F	E	P	FG	E	GE	P	P	F	E

Table 4.8 - General Properties of Commonly Known Elastomers (Courtesy of Parker) 105

4.14- Sealing Solutions for Hydraulic Cylinders
4.14.1- Considerations for Hydraulic Cylinders Reliable Sealing
❖ **For a cylinder to perform reliably under leak-free conditions →**
- Seals must be able to function at specified pressure and temperature.
- Seals must be able to withstand expected pressure spikes.
- Seals must be able to carry expected lateral loads.
- Seals must be compatible with type of hydraulic fluid used.
- Seals must offer conditions of low friction.
- Seals must be designed for easy installation and port passing.

❖ **To help cylinder seals to perform reliably:**
- Cylinder must be mounted coaxially with the load → lateral force ↓.
- Cylinder mounting points should be attached to a non-vibrating frame.
- Cylinder should operate in a clean environment.
- Cylinder should operate within the recommended working T & P.

❖ **Sealing Solution:** There is no one sealing solution that is good for all cylinders.

106

106

CYLINDER SPECIFICATION		LIGHT-DUTY		MEDIUM-DUTY		HEAVY-DUTY	
PRESSURE	Max	350 bar	5000 psi	500 bar	7500 psi	700 bar	10000 psi
	Normal Working	160 bar	2300 psi	250 bar	3625 psi	400 bar	5800 psi
		No pressure peaks		Intermittent pressure peaks		Regular pressure peaks	
Design		Lower operating stresses. Rigid well- aligned mounting, minimal side loading.		Steady operating stresses with intermittent high stress, some side loading.		Highly stressed for the majority of its working life. Side loading common.	
Condition of Fluid		Good system filtration. No cylinder contamination likely.		Good system filtration, but some cylinder contamination likely.		Contamination unavoidable from internal and external sources.	
Working Environment		Clean and inside a building. Operating temperature variations limited.		Mixture of indoors and outdoors but some protection from the weather.		Outdoors all the time or dirty indoor area. Wide variations in temperature, both ambient and working. Difficult service conditions.	
Usage		Irregular with short section of stroke at working pressures. Regular usage but at low pressure.		Regular usage with most of the stroke at working pressure.		Large amount of usage at high pressure with peaks throughout the stroke.	
Typical Applications		Machine tools Lifting equipment Mechanical handling Injection moulding machines Control and robot equipment Agricultural machinery Packaging equipment Aircraft equipment Light duty tippers		Heavy duty lifting equipment Agricultural equipment Light duty off-road vehicles Cranes and lifting platforms Heavy duty machine tools Injection moulding machines Some auxiliary mining machinery Aircraft equipment Presses Heavy duty tippers (telescopic) Heavy duty mechanical handling		Foundry and metal fabrication plant Mining machinery Roof supports Heavy duty earthmoving machinery Heavy duty off-road vehicles Heavy duty presses	

Equipment Manufacturers are the best source of information

Table 4.9 – Working Conditions and Typical applications for Hydraulic Cylinders (Courtesy of Hallite Seals)

107

107

201/335

❖ **Basic Components of Light-Duty Cylinder Sealing Solution:**
▪ Piston Seal Package:
  o Piston dynamic seal and a wear or a guide ring.
  o Piston static seal (if the piston head is assembled with the piston rod).

▪ Rod Seal Package:
  o Rod dynamic seal and a Guide Ring.
  o Rod Wiper and a Buffer Seal.
  o Rod static seal (seal the cylinder head against the barrel to prevent external leakage).

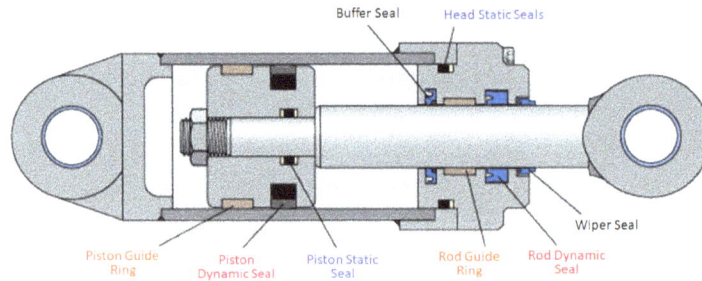

Fig. 4.70 - Basic Components of Cylinder Sealing Solutions   108

108

4.14.2- Sealing Solutions for Cylinder Rods

4.14.3- Sealing Solutions for Cylinder Pistons

**Examples of Sealing Solutions
for Cylinder Rods and Pistons
review in textbook:**

▪ **Figure 4.71 through 4.100**

109

109

4.14.4- Piston and Rod Design for Proper Sealing

- Designs for proper sealing functions:
  - Design of Seal Groove.
  - Design of cylinder piton and rod Lead-in Chamfers.
  - Design of Extrusion Gap.
  - Design of Mating Surface Finish.

4.14.4.1- Design of Seal Groove and Lead-In Chamfers

- Eliminate sharp edges by proper rounding.
- Improper lead-in chamfers → tearing of the O-Ring during installation.
- Seals manufacturers → instructions for lead-in chamfers based on seal size.

Fig. 4.101 - Improper Lead-in Chamfer Cause Tearing of an O-Ring During Installation (Courtesy of Trelleborg)

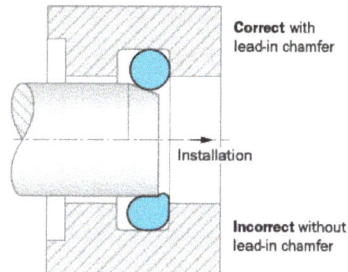

110

TECHNICAL DETAILS

OPERATING CONDITIONS	METRIC	INCH
Maximum Speed	1.0 m/sec	3.0 ft/sec
Temperature Range	-30°C +110°C	-22°F +230°F
Maximum Pressure	250 bar	3600 psi

CHAMFERS & RADII				
Groove Section ≤S mm	3.75	5.50	7.75	10.50
Min Chamfer C mm	2.00	2.50	5.00	5.00
Max Fillet Rad r₁ mm	0.40	0.80	1.20	1.60
Groove Section ≤ S in	0.150	0.220	0.310	0.410
Min Chamfer C in	0.080	0.100	0.200	0.200
Max Fillet Rad r₁ in	0.016	0.032	0.047	0.063

Fig. 4.102 - Data Sheet for a Double Acting Piston Seal 764 with Pre-Loaded O-Ring (Courtesy of Hallite Seals)

111

4.14.4.2- Extrusion Gap Design

Example 1: Use of graphs provided by the manufacturer:
- Diametral Clearance (DC) means Extrusion Gap.

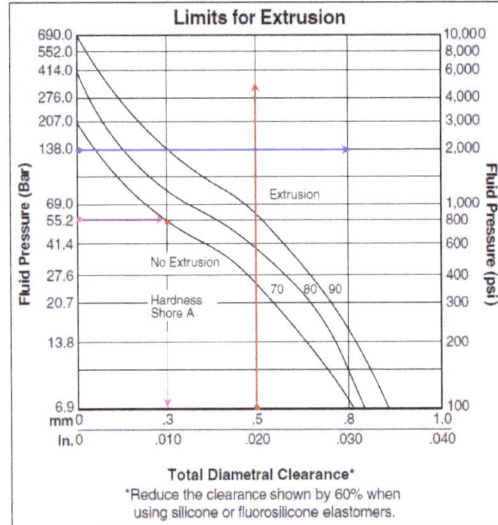

- **Softer seals extrude at low P**
- At same DC
- → P ↑
- → seal hardness should ↑

- At same P
- → seal hardness ↓
- → required DC ↓

- At same hardness
- → P ↑
- → required DC ↓

Fig. 4.103 - Limits of Extrusion
(Courtesy of Parker)   112

112

Case Study:
- For fluid pressure of 55.2 bar (800 psi).
- Seal hardness 70 Shore A.
- → maximum allowable gap extrusion is 0.3 mm (0.01 in).

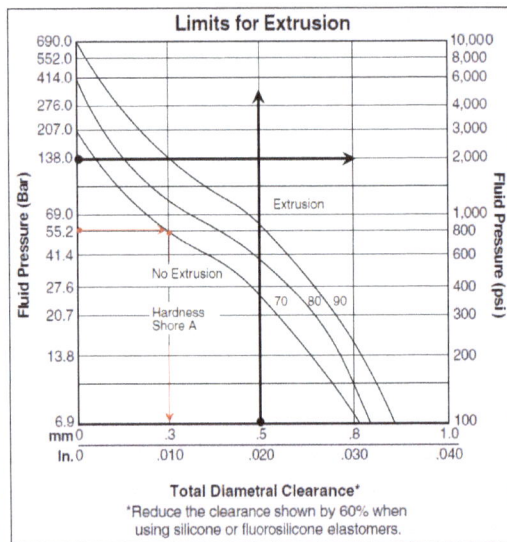

113

113

Example 2- Use of dimensional parameters and a nomograph provided by the manufacturer:

e = Maximum sealing and anti-extrusion gap.
D = Piton diameter.
d = Rod diameter.
S = Cross section.
P = Working pressure
T = Working Temperature.

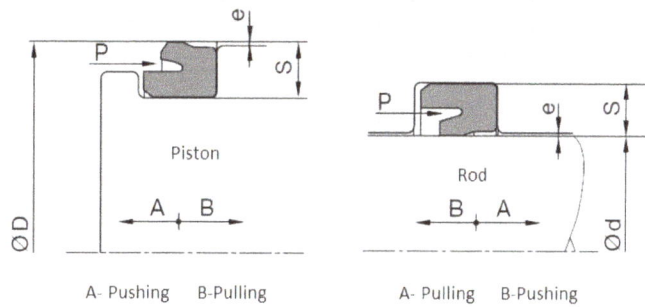

Fig. 4.104 - Anti-Extrusion Gap (Courtesy of Parker)

114

Case Study:
- d/D = 100 mm, S = 6 mm, P = 100 bar, and T = 80 OC → Sealing Gap e = 0.18 mm

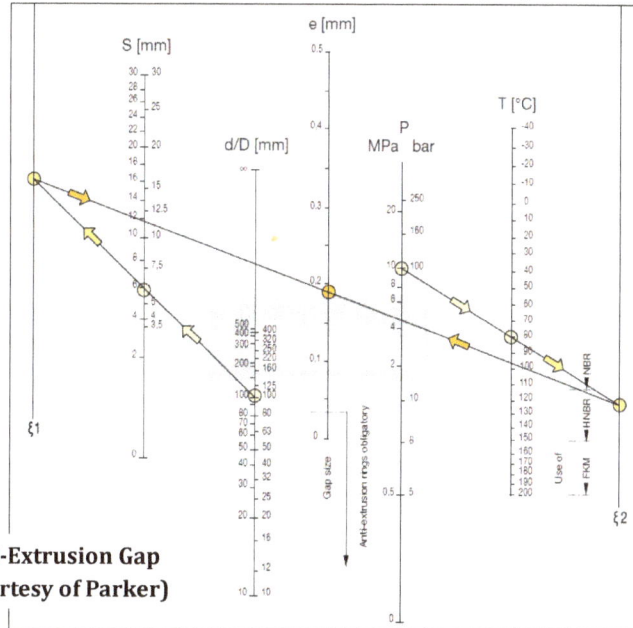

Fig. 4.105 - Anti-Extrusion Gap Nomograph (Courtesy of Parker)

115

Example 3- Use of tabulated results provided by the manufacturer:

MAXIMUM EXTRUSION GAP			
Pressure bar	100	160	250
Maximum Gap mm	0.60	0.50	0.40
Pressure psi	1500	2400	3750
Maximum Gap in	0.024	0.020	0.016

**Table 4.10 - Data Sheet for a Double Acting Piston Seal
764 with Pre-Loaded O-Ring
(Courtesy of Hallite Seals)**

116

116

4.14.4.3- Design of Mating Surface Finish

❖ **Dynamic Seals:**
- Too Smooth → improper retaining of lubrication film
- → frictional heat → excessive seal wear.
- Too Coarse → small cuts or scores in the sealing lip
- → premature seal failure.

❖ **Static Seals:**
- Static sealing surface finish must not be ignored.
- Static surfaces free from chatter marks → better control of leakage.

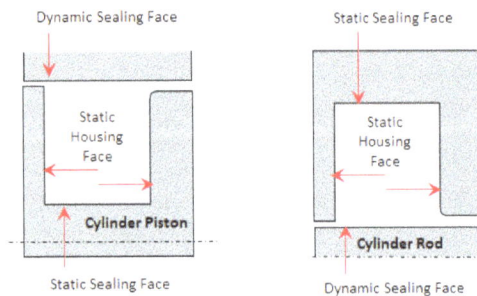

Fig. 4.106 - Dynamic and Static Sealing Surfaces in a Hydraulic Cylinder

117

117

- Most common surface finish measurements in the fluid power industry.
- **ISO 4287** and **ISO 4288**.

- **Case 1 (Ra):**
- Arithmetical mean deviation of an absolute ordinate over the evaluation length.

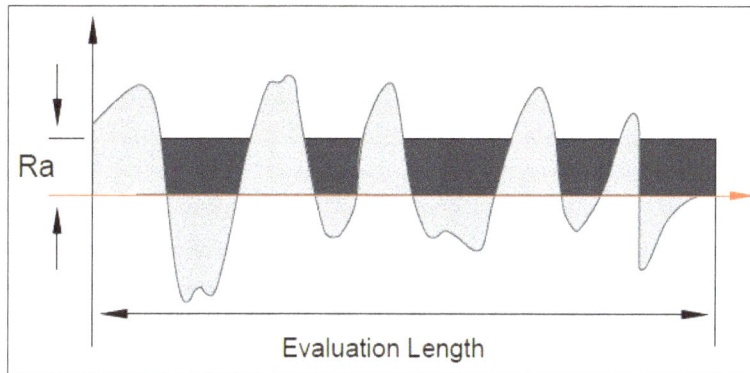

Fig. 4.107 - Surface Finish Measurement Ra (Courtesy of Hallite Seals)

118

118

- **Case 2 (Rt):**
- Sum of height of the largest profile peak height **Rp** and the largest profile valley **Rv** over the evaluation length.

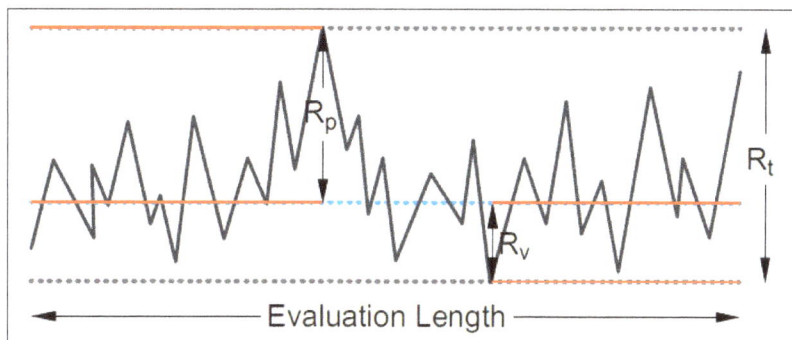

**Fig. 4.108 - Surface Finish Measurement Rt
(Courtesy of Hallite Seals)**

119

119

- **Case 3 (Rz):**
- **Rz(n):** Sum of height of the largest profile peak height **Rp** and the largest profile valley **Rv** within a sampling length.
- **Rz** is then the average of **Rz(n)** over the evaluated length.

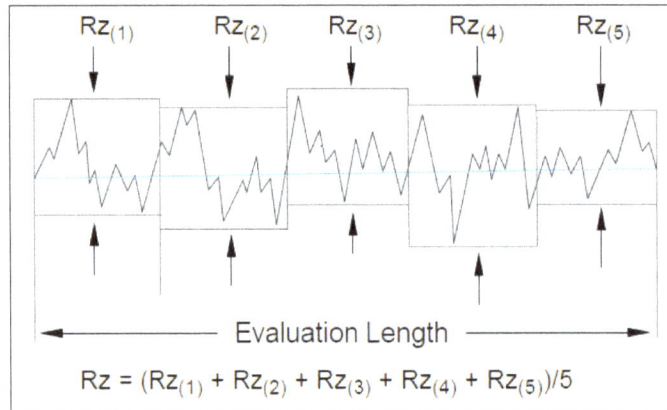

Fig. 4.109 - Surface Finish Measurement R_Z
(Courtesy of Hallite Seals)

120

SURFACE ROUGHNESS	µmRa	µmRz	µmRt	µinRa	µinRz	µinRt
Dynamic Sealing Face ØD₁	0.1 - 0.4	1.6 max	4 max	4 - 16	63 max	157 max
Static Sealing Face Ød₁	1.6 max	6.3 max	10 max	63 max	250 max	394 max
Static Housing Faces L₁	3.2 max	10 max	16 max	125 max	394 max	630 max

**Table 4.11 - Data Sheet for a Double Acting Piston Seal
764 with Pre-Loaded O-Ring
(Courtesy of Hallite Seals)**

121

4.15- Sealing Solutions for Rotational Shafts

- Prevent external leakage in pumps, motors, and rotary actuators.
- Axial, radial, or combined sealing directions.
- Working conditions and applications → seal materials and design.

**Fig. 4.110 - Various Materials and Designs for Rotational Shaft Seals
(Courtesy of American High-Performance Seals)**

122

122

4.15.1- Rotational Radial Seals

- A Radial Seal acts radially against the shaft → prevents axial leakage.

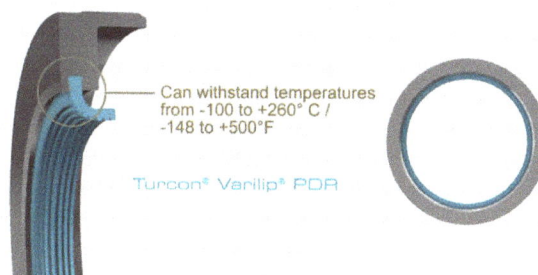

Can withstand temperatures from -100 to +260° C / -148 to +500°F

Turcon® Varilip® PDR

Fig. 4.111 – Light Duty Rubber Lip Rotational Radial Seal

Video 314 (3.0 min)

123

123

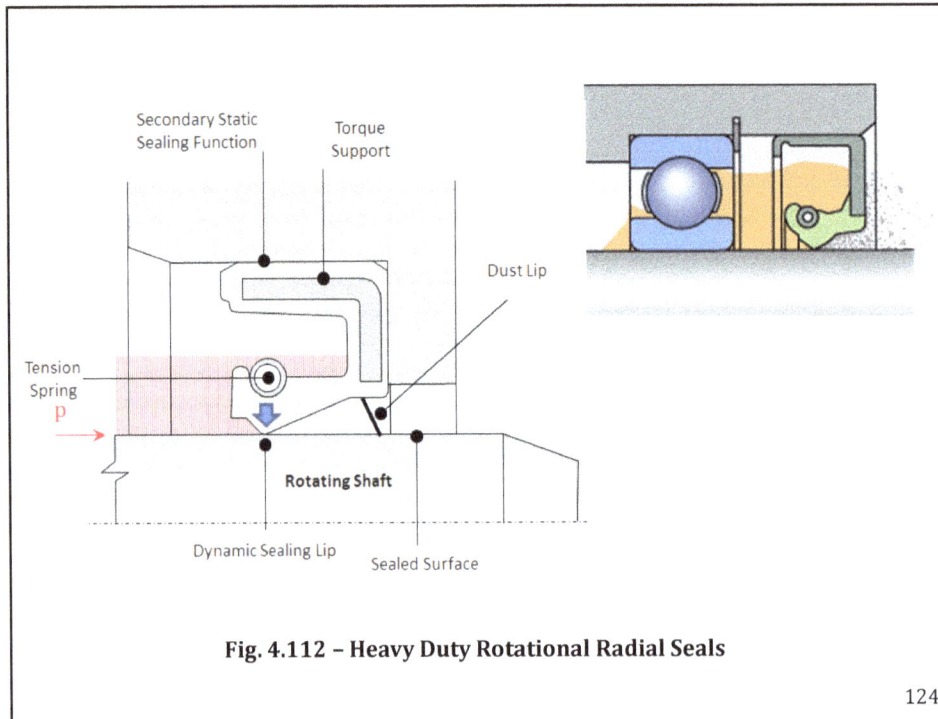

Fig. 4.112 – Heavy Duty Rotational Radial Seals

124

124

- **Pressure:** Available for wide range of P 1 to 700 bar (15 to 10152 psi).
- **Designs:** Supporting inserts, springs, and sealing lips.
- **Applications:** power transmissions, motors, pumps, gearboxes, fans, etc.

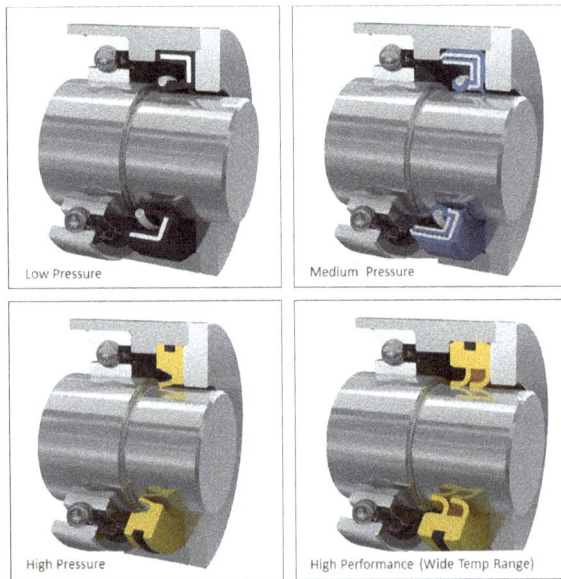

Video 315 (3.0 min)

Video 316 (1.5 min)

Fig. 4.113 - Constructional Details of the Most Common Rotational Radial Seals (Courtesy of American High-Performance Seals)

125

125

- ▪ Waved rubber on the static sealing surface →
- ▪ Improved seal retaining force in the groove in the housing →
- ▪ Prevent seal rotation in its groove.

	With Rubber Case		With Metallic Case
Flat Surface		DIN Type B without Dust Lip without Support	
DIN Type A Waved Surface without Dust Lip		DIN Type BS with Dust Lip without Support	
DIN Type AS Waved Surface with Dust Lip		DIN Type C without Dust Lip with Support	
		DIN Type CS with Dust Lip with Support	

Fig. 4.114 - Configurations of Rotational Radial Seals to DIN 3760/ISO 6194 Standards

126

126

4.15.2- Rotational Axial Seals

❖ **V-Ring Rotational Axial Seals:**

- ▪ An Axial Seal rotates with the shaft → prevents radial leakage.
- ▪ This seal is stretched and mounted directly on the shaft.
- ▪ It is held on the shaft by the inherent tension of the rubber body.
- ▪ Used for a broad range of applications.
- ▪ Used as a secondary seal to protect a primary seal.
- ▪ Flexible sealing lip → light contact pressure against the surface.

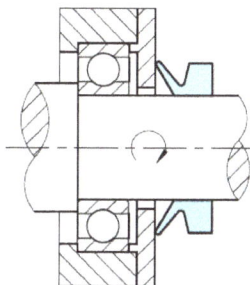

Fig. 4.115 - Classic Rotational Axial Seal (Courtesy of Trelleborg)

127

127

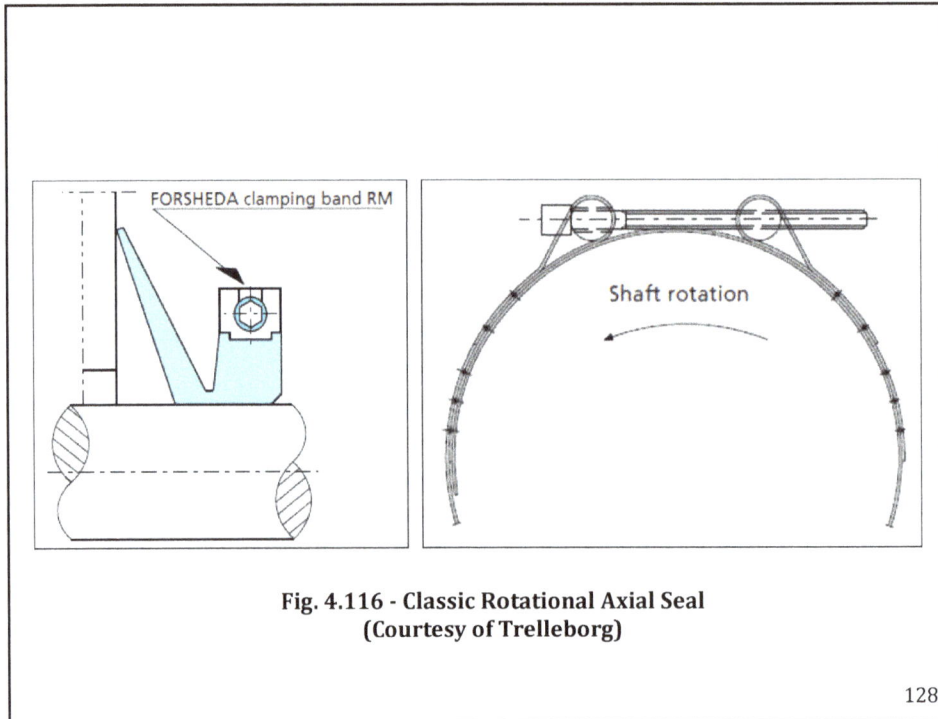

Fig. 4.116 - Classic Rotational Axial Seal
(Courtesy of Trelleborg)

128

128

❖ **Power Losses in Rotational Shafts**:

- Test Conditions: Two types of V-Rings in dry-running condition.

- Contact Surface: Unhardened steel surface.

- Test Results:

- Peripheral (tangential) speed (m/s) and shaft diameters (d in mm).

- Peripheral Speed ↑→ centrifugal force ↑→

- Contact pressure of the sealing lip decreases ↓→

- Frictional losses and heat are ↓ →

- Excellent wear characteristics and extended seal life.

- In the 15 - 20 m/s range the friction reduces to zero.

129

129

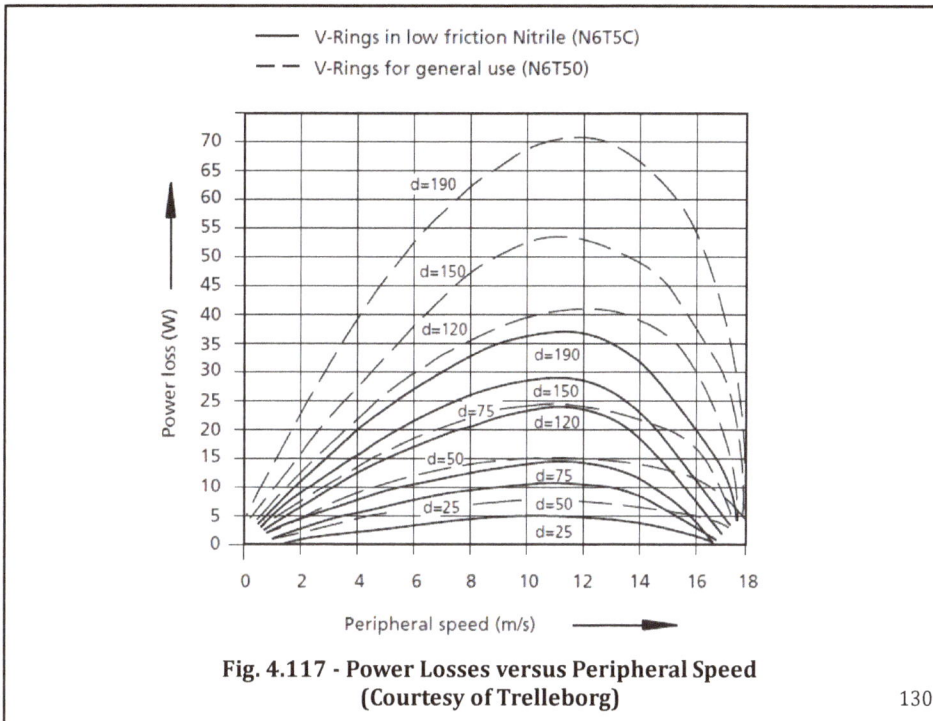

**Fig. 4.117 - Power Losses versus Peripheral Speed
(Courtesy of Trelleborg)**

130

130

❖ **GAMMA Rotational Axial Seals:**
- Enhanced design consists of (sealing element + metal case).
- → sealing against foreign matter, liquid splatter, and grease.

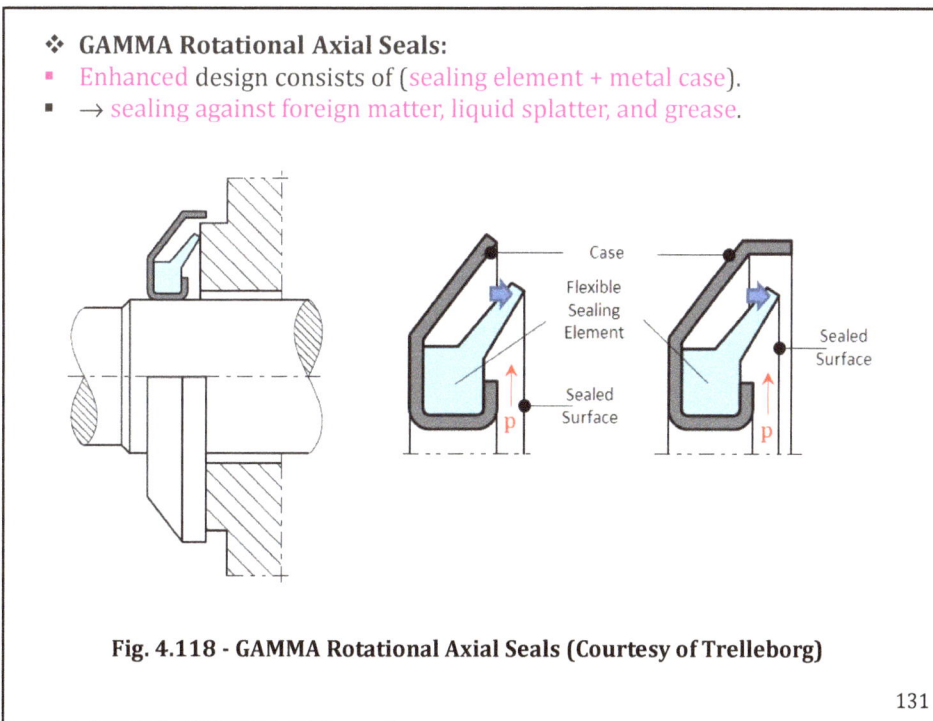

Fig. 4.118 - GAMMA Rotational Axial Seals (Courtesy of Trelleborg)

131

131

❖ **Enhanced Axial Shaft Seals:**
- Are used primarily as a protective seal for bearings.
- Their sizes are matched to those of roller bearings.
- Internal seal lip → prevent fluid from escaping.
- External seal lip → sealing grease and dirt entering from the outside.
- Both types consist of:
    o An elastomer-elastic membrane with axial sealing lip.
    o A metallic reinforcement ring acting axially.

**Fig. 4.119 - Enhance Rotational Axial Seals
(Courtesy of Trelleborg)**

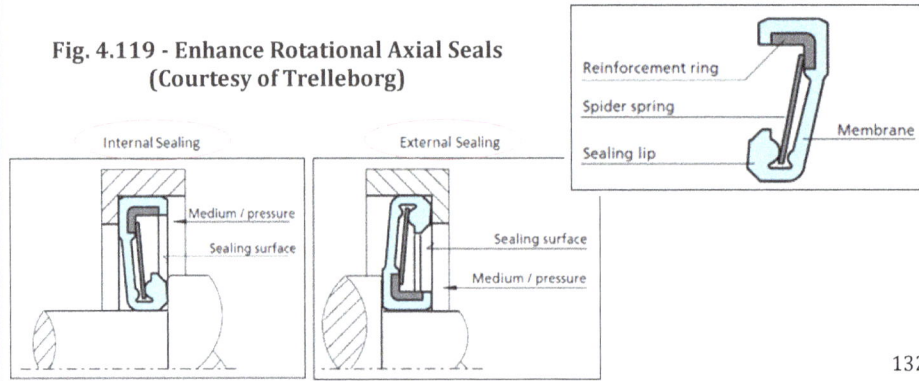

132

132

4.15.3- Combined Axial/Radial Sealing Solutions for Rotational Shafts

- Primary Seal: A flexible rubber V-Ring rotational axial seal.
- Secondary Seal: A spring-energized rotational radial seal.

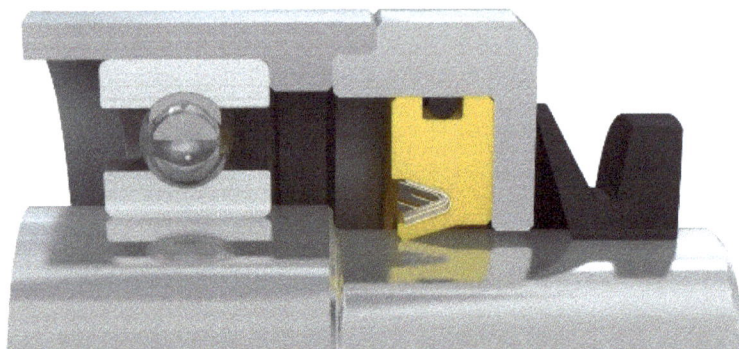

**Fig. 4.120 - Combined Axial/Radial Sealing Solutions for Rotational Shafts
(Courtesy of American High-Performance Seals)**

133

133

Chapter 4 Reviews

1. Among the factors that affect the lifetime of hydraulic seals are?
   A. Humidity, vibration and contamination.
   B. Working temperature and pressure.
   C. Surface finish and geometry of the seal.
   D. All the above.

2. Cylinder rod seal is considered a?
   A. Static seal.
   B. Dynamic rotational seal.
   C. Dynamic translational seal.
   D. None of the above.

3. Pump shaft seal is considered a?
   A. Static seal.
   B. Dynamic rotational seal.
   C. Dynamic translational seal.
   D. None of the above.

4. Seal hardness is measured by the following scale:
   A. PSI.
   B. GMP.
   C. BAR.
   D. SHORE A.

5. Wear Rings are basically used to:
   A. Offer bearing surface to absorb lateral forces.
   B. Seal rod leakage in one direction.
   C. Remove dust and dirt from the shaft when the cylinder retracts.
   D. Prevent seal extrusion.

6. Backup Rings are basically used to:
   A. Offer bearing surface to absorb lateral forces.
   B. Seal rod leakage in one direction.
   C. Remove dust and dirt from the shaft when the cylinder retracts.
   D. Prevent seal extrusion.

7. Wipers are basically used to:
   A. Offer bearing surface to absorb lateral forces.
   B. Seal rod leakage in one direction.
   C. Remove dust and dirt from the shaft when the cylinder retracts.
   D. Prevent seal extrusion.

8. The hydraulic seal property "Resilience" means:
   A. The ability of a compound to return quickly to its original shape after a temporary deflection Seal rod leakage in one direction.
   B. The percentage increase in length with respect to the original length
   C. The ratio of the force achieved at the moment of rupture and the initial cross-section thickness of the specimen
   D. the volumetric decrease of the seal after it has been in contact with a fluid medium.

9. The hydraulic seal property "Shrinkage" means:
   A. The ability of a compound to return quickly to its original shape after a temporary deflection Seal rod leakage in one direction.
   B. The percentage increase in length with respect to the original length
   C. The ratio of the force achieved at the moment of rupture and the initial cross-section thickness of the specimen
   D. the volumetric decrease of the seal after it has been in contact with a fluid medium.

10. The hydraulic seal property "Tear Strength" means:
    A. The ability of a compound to return quickly to its original shape after a temporary deflection Seal rod leakage in one direction.
    B. The percentage increase in length with respect to the original length
    C. The ratio of the force achieved at the moment of rupture and the initial cross-section thickness of the specimen
    D. the volumetric decrease of the seal after it has been in contact with a fluid medium.

Chapter 4 Assignment

Student Name: --- Student ID: ------------------

Date: -- Score: ------------------------

Assignment: In the shown below compression test, the following dimensions have been measured. Calculate the compression set for this seal material

- h_o = 2 cm
- h_s = 1.8 cm
- h_i = 1.6

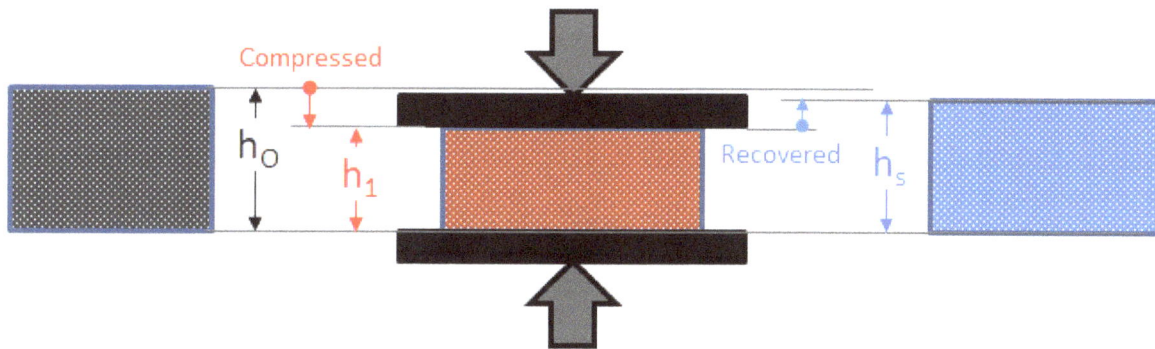

Chapter 5
Hydraulic Heat Exchangers

Objectives:
This chapter overviews various types of heat exchangers including air-type, water-type, and plate-type. Construction, operation, features, applications, and sizing calculations are discussed.

Brief Contents:
5.1- Contribution of Heat Exchangers:
5.2- Air-Type versus Water-Type Oil Coolers
5.3- Determination of Cooling Capacity for an Oil Cooler
5.4- Air-Type Oil Coolers
5.5- Shell-and-Tube Water-Type Oil Coolers
5.6- Plat-Type Oil Coolers
5.7- Cooling-Filtration Units
5.8- Oil Cooling Circuit Diagram
5.9- Oil Temperature Automatic Control Solutions
5.10- Electrical Oil Heaters

0

0

The following topics are discussed in Chapter 11 in Volume 5
"Maintenance and Safety"

- BP-Heat Exchangers-01-Selection and Replacement

- BP-Heat Exchangers-02-Maintenance Scheduling

- BP-Heat Exchangers-03-Installation and Maintenance

- BP-Heat Exchangers-04-Standard Tests and Calibration

The following topics are discussed in Chapter 11 in Volume 6
"Troubleshooting and Failure Analysis"

- Heat Exchangers Inspection

- Heat Exchangers Troubleshooting

- Heat Exchangers Failure Analysis

1

1

5.1- Contribution of Heat Exchangers

Contribution:

- Removing or adding heat to the hydraulic fluid →
- Maintain the change in temperature within acceptable limits.

- Working T affects:
  - Fluid viscosity and other properties.
  - Fluid degradation and service lifetime.
  - Performance of hydraulic fluid additives.
  - Machine startup in cold weather.
  - Hydraulic components lubrication.
  - Hydraulic components wear and lifetime.
  - Pressure drop through the components and transmission lines.
  - Overall system energy efficiency.

2

2

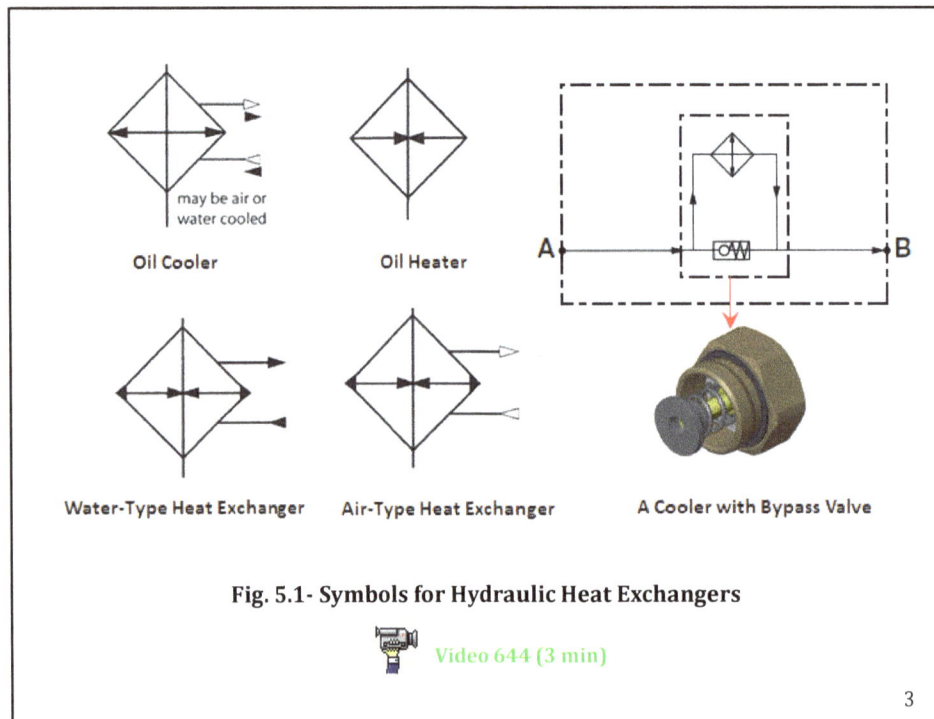

Oil Cooler (may be air or water cooled)

Oil Heater

A ——— B

A Cooler with Bypass Valve

Water-Type Heat Exchanger

Air-Type Heat Exchanger

Fig. 5.1- Symbols for Hydraulic Heat Exchangers

Video 644 (3 min)

3

3

5.2- Air-Type versus Water-Type Oil Coolers

❖ **Heat Removal:**

- <u>Passively:</u> → heat dissipation from the reservoir, transmission line, and other components.

- <u>Actively</u> by heat transfer using air-type or water-type oil coolers.

Air-Type Oil Cooler   Water-Type Oil Cooler

Fig. 5.2- Air-Type versus Water-Type Oil Coolers

4

4

❖ **Application:**
- Water-type oil coolers → commonly used in industrial applications (cooling water supply is available)
- Air-type oil coolers → commonly used for mobile applications.

❖ **Efficiency:**
- Water-type oil coolers → more efficient at higher ambient temperature.

 Video 640 (3 min)

❖ **Size:**
- Water-type coolers → higher cooling density:
  o Higher cooling capacity for same physical size
  o OR same cooling capacity for less physical size

❖ **Noise:**
- Water-type oil coolers → Less noise (no fan).
- Air-type cooler → typical noise level is 60-90 db.

5

5

❖ **Maintenance:**
- Frequent cleaning is required for both coolers, especially when:
  o Water-type oil cooler → sea water are used.
  o Air-type oil cooler → work environments is highly contaminated.

❖ **Fluid Compatibility:**
- Aggressive type of oil is used →
- Consult cooler manufacturer (regardless the type of the cooler)

- For water-type oil coolers:
  o De-mineralized and untreated water → may be used without concern.
  o Sea water or chemically treated water → consult the manufacturer.

6

6

❖ **General Heat Transfer Terminologies:**

- In general, heat transfer could occur by any of the following:

- **Conduction:** Heat transfer between solid bodies.

- **Convection:** Heat transfer to or from fluids (liquid or gas).

- **Radiation:** Heat transferred through vacuum in straight lines from hot spot to cold surrounding.

7

7

❖ **Heat Transfer in Water-Type Oil Coolers:**
▪ Water passes in the tubes and hot oil pass in the shell →
1. Convection (hot oil → outer surface of the tube wall).
▪ Turbulent flow → improved heat transfer rate.

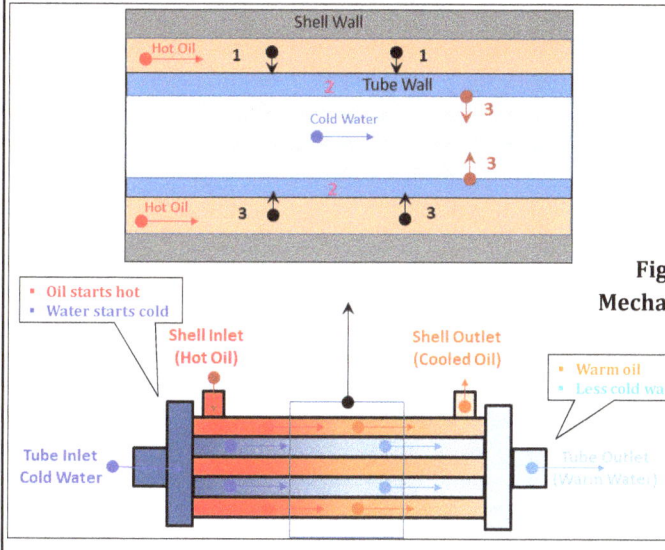

Fig. 5.3 - Heat Transfer Mechanisms in in Water-Type Oil Coolers

8

2. Conduction (through the tube wall).
▪ Tubes made from copper or aluminum alloys →
▪ high thermal conductivity → improved heat transfer rate.

Fig. 5.3

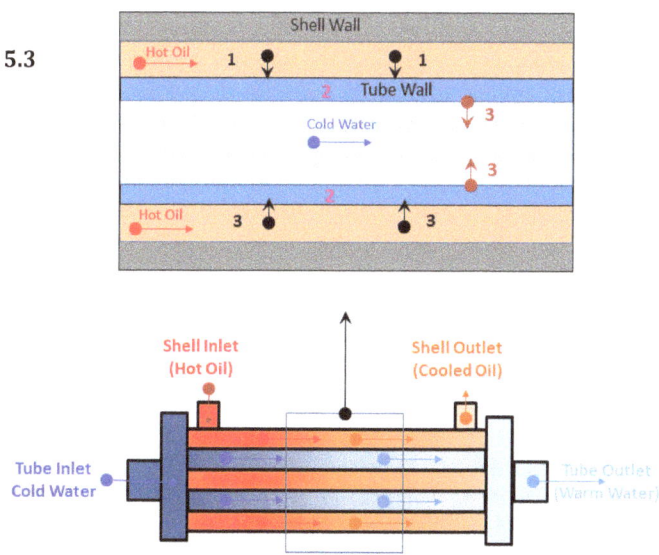

9

3. Convection (tube inner surface → cooling water in the tube).
○ Use of multi-pass and turbulent flow → improved heat transfer values.

Fig. 5.3

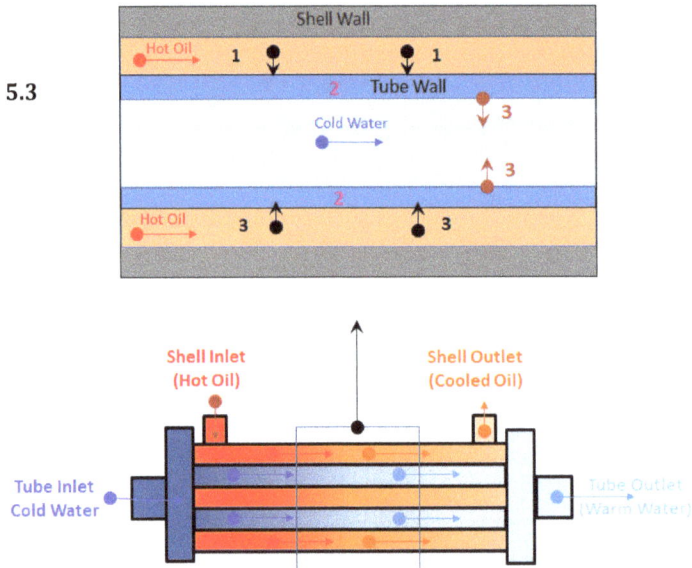

10

❖ **Heat Transfer in Air-Type Oil Coolers:**
1. Convection (hot oil → the inner surface of the tube wall and fins).
▪ Turbulent flow → improved heat transfer.
2. Conduction (through the tube wall).
3. Convection (outer surfaces of the tubes and fins → cooling air.

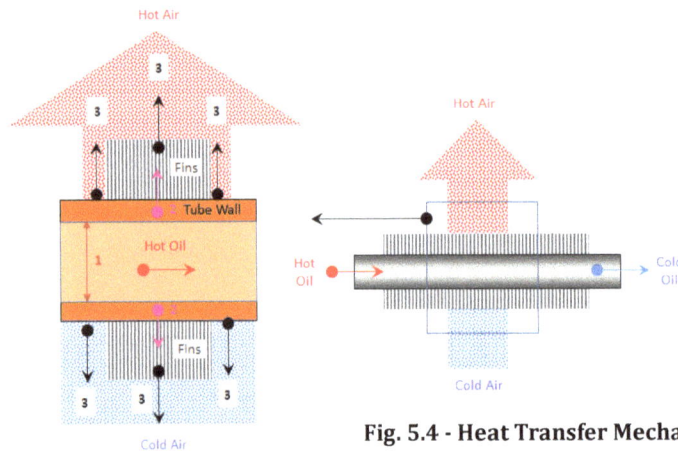

**Fig. 5.4 - Heat Transfer Mechanisms in
Air-Type Oil Coolers**

11

5.3- Determination of Cooling Capacity for an Oil Cooler

❖ **Basic Concepts:**
- Power added to hydraulic system → no useful work →
- wasted energy → heat generation.

- Determination of the Cooling Capacity →
- sizing and selection of oil cooler.

❖ **Approximate Estimation of Cooling Capacity:**
- Hydraulic systems are approximately 75% efficient → Rule of thumb →
- Cooling capacity = ¼ of the max power (check all phases of operation).

Example: A hydraulic driven die casting molding machine has a 5-phases duty cycle. Maximum power required to drive this machine = 20 HP, then:

Cooling Capacity q = 20/4 = 5 HP = 5 HP x 2,545 = 12725 BTU/HR
(2,545 BTU/HR = 1 horsepower)

12

12

❖ **Precise Calculation of Cooling Capacity:**
- Thermal Balance Equation

Cooling Capacity q = Total Heat Gained – Total Heat Dissipated   5.1

- Sources of Gaining Heat:
  o Internally: Power losses in hydraulic components and transmission lines.
  o Externally: Sources (sun, furnace, etc.)

- Sources of Heat Dissipation:
  o Through surfaces of components, lines, and walls of the reservoir.
  o Note: Only heat dissipated through the reservoir walls can be considered for safe sizing of the cooler.

13

13

Example:

A thermal analysis of a hydraulic system is conducted and resulted in:

Losses:

- Fixed displacement pump inefficiency → losses = 10 HP.
- Flow control valve (15 gpm at ΔP = 750 psi) → losses = 6.5HP.
- Directional control valve (5 gpm at ΔP = 100 psi) → losses = 2.5HP.
- Hydraulic tubing (30 gpm at ΔP = 30 psi) → losses = 1HP.
- Total Power Losses = 10 + 6.5 + 2.5 + 1 = 20 HP.
- Total Heat Gained = 20 HP x 2,545 = 50,900 BTU/HR.

Total Heat Dissipated from Reservoir:

- = 5 HP = 5 HP x 2,545 = 12725 BTU/HR.

Thermal Balance:

- Eq. 5.1→ Cooling Capacity q = 50,900 – 12725 = 38,175 BTU/HR.

14

14

❖ **Oil Cooler Sizing:**

$$A\left(ft^2\right) = \frac{q\,(BTU/HR)}{\Delta T(^oF) \times U\left[\frac{BTU/HR}{^oF \times ft^2}\right]}$$

5.2

- **A (ft^2)** = Heat transfer surface area of the cooler.
- **q (BTU/HR)** = Cooling capacity required by the cooler.
- **ΔT (^oF)** = Temperature difference of fluids at the entrance of the cooler.
- **U [(BTU/HR)/(^oFxft^2)]** = *Coefficient of heat transfer* of the cooler.

Example:

- Cooling capacity required for a hydraulic system = 9,000 BTU/HR.
- ΔT (between hot oil and cold water) = 30 ^oF.
- Cooler coefficient of heat transfer = 100 [(BTU/HR)/(^oF x ft^2)].
- Eq. 5.2 → A (ft^2) = 9000 / (30 x 100) = 3
- However, manufacturers provide selection charts.

15

15

5.4- Air-Type Oil Coolers
5.4.1- Construction and Operation of Air-Type Oil Coolers

❖ **Construction:** Cooling Fan and Radiator.

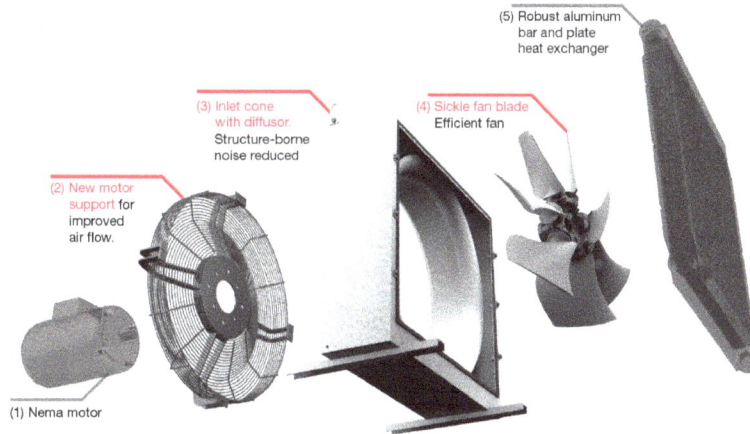

Fig. 5.5- Construction of Air-Type Oil Coolers (Courtesy of Hydac)

16

16

❖ **Operation:**

- Hot oil is circulated through the radiator.
- A fan blow air through the fins of the radiator..

❖ **Heat Transfer:**

- Convection (hot oil → inner surface of tubes walls).
  - Radiator design → Devices placed inside the tubes
  - → generate turbulence → heat transfer ↑ BUT pressure drop ↑

- Conduction (through the walls of the tubes and fins).
- Convection (outer surfaces of tubes and fins → flow of forced air).

17

17

❖ **Driving Motor:** A cooling fan can be driven by an AC or a DC electric
   motor. It can also be driven by a hydraulic motor.

**Fig. 5.6- Driving Motors for Air-Type Oil Coolers
(Courtesy of Hydac)**

18

18

5.4.2- Sizing of Air-Type Oil Cooler

❖ Selection of an air Cooler of a specific Size is function of:

▪ Oil flow through the cooler.

▪ Oil Viscosity.

▪ EDT is Entering ΔT (hot oil - ambient air).

▪ Example:

o A hydraulic system requires cooling capacity q = 30,000 BTU/HR.

o Oil Viscosity = 250 SSU

o Entering Temperature Difference (EDT) between oil and air = 30 $^\circ$F.

o → Normalized Cooling Capacity

o q = 30,000 (BTU/HR) / 30 $^\circ$F = 1000 (BTU/HR)/$^\circ$F.

o Oil Flow through the cooler is 40 gpm.

19

- o **Step 1 (Size):**
- o Cooling performance graph → locate the operating point →
- o Closest cooler is "ULAC-016B" → cooling capacity ≈ 1100 (BTU/hr)/$^{\circ}$F
- o → The cooler is ≈ 10% larger than needed.

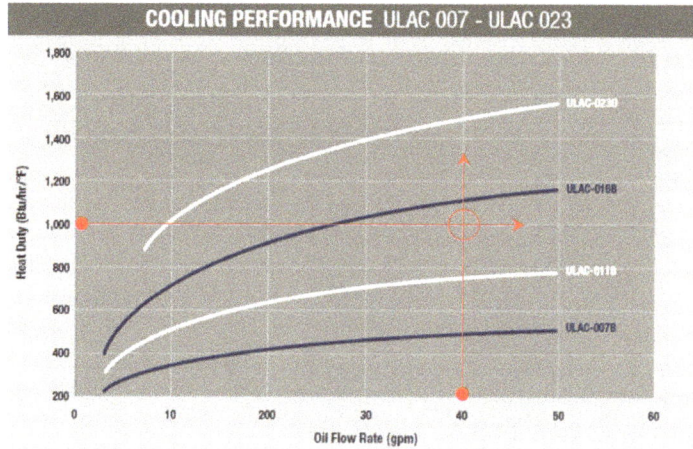

Fig. 5.7 - Cooling Performance of Air-Type Oil Coolers (Courtesy of Parker) 20

20

- o **Step 2 (Pressure Drop):**
- o Oil flow = 40 gpm → pressure drop = 18 psi.
- o Pressure Drop correction (based on oil Viscosity = 250 SSU)
- o = 18 x 1.5 = 27 psi.

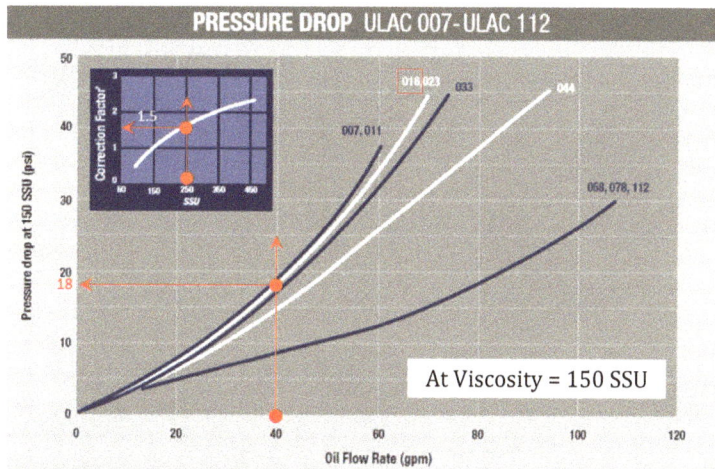

Fig. 5.8- Pressure Drop Across Air-Type Oil Coolers (Courtesy of Parker) 21

21

5.5- Shell-and-Tube Water-Type Oil Coolers

5.5.1- Construction and Operation of Shell-and-Tube Oil Coolers

❖ **Construction (Multi-Pass & Counter-Flow):**

▪ Bundle of tubes enclosed in a metal shell.

▪ Tubes are made of material that has high thermal conductivity.

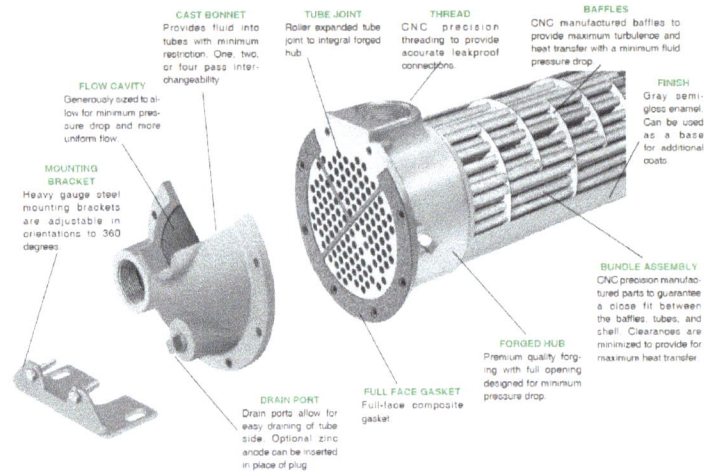

Fig. 5.9- Construction of Shell-and-Tube Oil Coolers (www.SouthwestThermal.com)   22

22

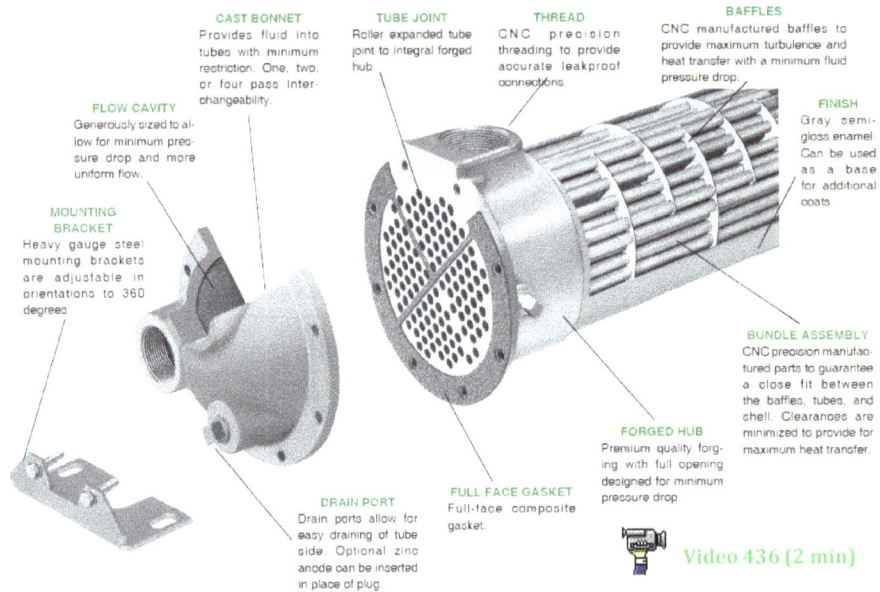

Video 436 (2 min)

❖ **Leakage:** from water to oil or vise versa depends on which has higher pressure.

23

❖ **Operation:**
- **Why** Hot oil flows in the shell and cooling water flows in the tube?:
○ Promote Heat dissipation to surroundings.
○ Water rust and contamination → easy tube bundle removal and cleaning.
○ Thick shell walls side → tolerates P and T.
○ If cold water is in the shell → condensation or humidity on the outer surface → water ingression into the system.
○ If cold water is in the shell → water is electrical conductor →
○ Galvanic corrosion between the shell metal and the tube metal.

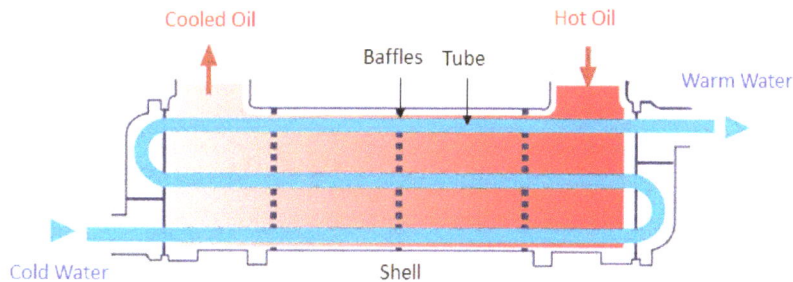

Fig. 5.10- Basic Construction of Shell-and-Tube Oil Coolers   24

24

❖ **Heat Transfer:**
1. Heat is convected from the hot oil to the tube walls,
2. Heat is conducted through the tube wall,
3. Heat is convected to the cooling water in the tube.

❖ **Oil-Water Flow:**
- The following rules of thumb shows better efficiency:
- Rule-of-thumb 1: Oil flows at a rate of 1 m/s (3 feet/s) or less.
- Rule-of-thumb 2: Water flows ≈ one-half of the oil flow rate.
- This means the oil/water flow ratio is 2:1.

25

25

❖ **Galvanic Corrosion:**

- Some applications (e.g., marine & offshore) →

- Sea water is used as cooling water →

- If sea water flows in the shell (not in the tube) →

- Electrical current flow between metals of shell and tube →

- Ions move from one metal to other through the sea water →

- Galvanic Corrosion occur, if ignored →

- One metal is damaged, and the cooler fails →

- Zinc Anodes releases its electrons scarifying itself →

- Shell and tube metals are protected →

- Zinc anodes must be routinely inspected and replaced, if necessary.

26

26

Fig. 5.11- Zinc Anodes to Prevent Galvanic Corrosion
(https://ej-bowman.com/knowledge-centre/zinc-anodes)

Fig. 5.12- Zinc Anodes Inspection
(www.marinedieselbasics.com)

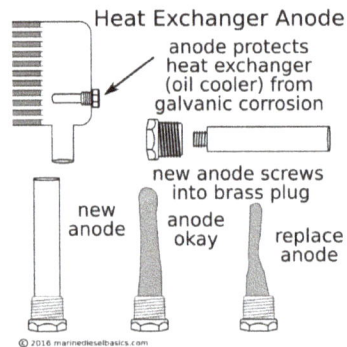

27

❖ **Number of Passes:**
- The number of times the cooling water travels through the length of the cooler between inlet and outlet ports.

- Single-Pass Shell and Tube Oil Coolers:
o Parallel or counter flow.

Fig. 5.13- Single-Pass Shell and Tube Oil Coolers

Video 166 (0.5 min)

28

- Multi-Pass Shell and Tube Oil Coolers:
o Two or four passes
o # of passes ↑→ better heat transfer and cooling capacity.
o Full parallel or counter flow is not possible.

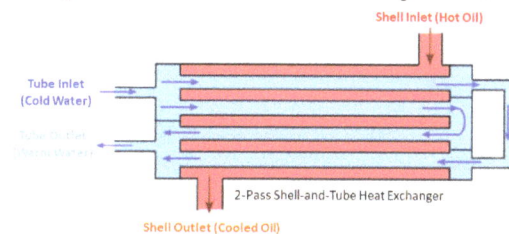

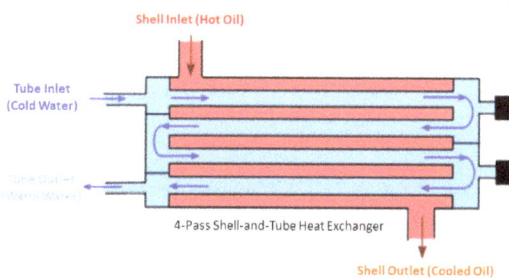

Fig. 5.14- Multi-Pass Shell and Tube Oil Coolers

29

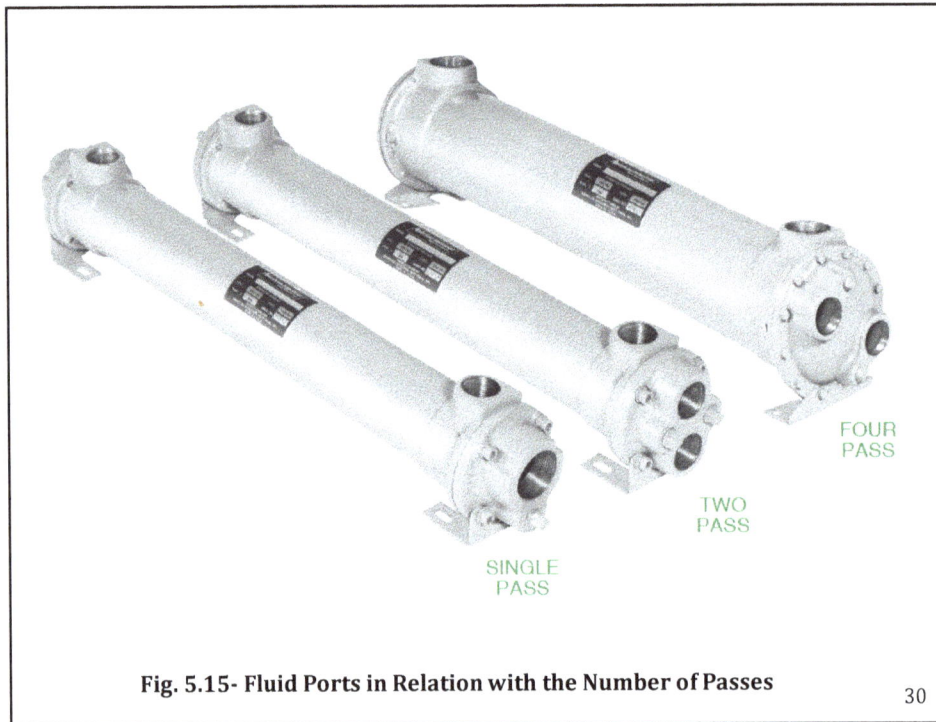

Fig. 5.15- Fluid Ports in Relation with the Number of Passes

30

30

Flow Direction:

1- Parallel-Flow:
o ΔT is highest at the inlet $\rightarrow$
o Nonhomogeneous thermal stress on the cooler parts.

Video 434 (0.5 min)

2- Counter-Flow:
o $\Delta T \approx$ constant $\rightarrow$
o Homogeneous thermal stress on the cooler parts $\rightarrow$
o Recommended

Animation 51

Fig. 5.16- Parallel-Flow versus Counter-Flow in Single-Pass Shell-and-Tube Oil Coolers

31

31

- Cross-Flow Shell-and-Tube Oil Coolers:
  - Oil flows through the shell perpendicularly on the tube →
  - Time for the hot oil to dissipate heat ↓→
  - Heat transfer efficiency ↓.

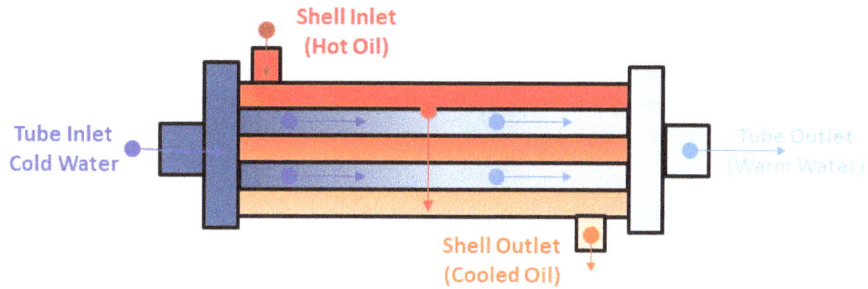

Fig. 5.17- Cross-Flow in Single-Pass Shell-and-Tube Oil Coolers

Video 435 (1 min)

32

32

5.5.2- Oil/Water Safety Heat Exchanger

- Avoid mixing the cooling water to oil.
- Two coaxial tubes.
- Cooling water → inner tube.
- Sealing liquid → outer tube.
- Sealing liquid surrounding the inner tube.
- Sealing liquid → high heat transfer.
- Accumulator →
  - Maintain sealing liquid pressure constant.
  - Compensates volumetric thermal change.
- P Switch → indicated tube leak.

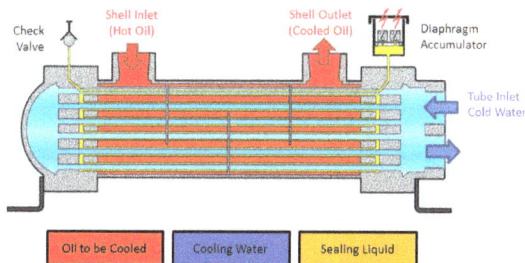

Fig. 5.18- Oil/Water Safety Heat Exchanger (universalhydraulik.com)

33

33

5.5.3- Sizing of Shell-and-Tube Water-Type Oil Coolers

**Fig. 5.19- Custom and Off-the-Shelf Shell and Tube Heat Exchangers
(www.fluiddynamics.com)**

❖ **Performance Curves:**
- Developed by manufacturer under the following conditions:
  - Oil Viscosity = 100 SUS.
  - Oil/Water Flow Ratio = 1:1 or 2:1 to match the curve results.
  - Design Entrance Temperature Difference, **DETD** = 40 $^\circ$F.
  - Oil Pressure Drop Codes (+ = 5 psi, ☆ = 10 psi, ○ = 20 psi, Δ= 50 psi).
- If a typical application has different conditions, correction must be made based on correction factors provided by the manufacturer.

34

34

Example:

❖ **Given Data:**
- Required Cooling Capacity q_r = 19.8 hp (based on system energy analysis).
- Oil Viscosity is 200 SUS (for equivalent ISO VG, refer to conversion tables).
- Oil Flow = 60 gpm.
- Actual Entrance Temperature Difference, **ETD** = 30 $^\circ$F
- Allowable Pressure Drop = 10 psi.

35

35

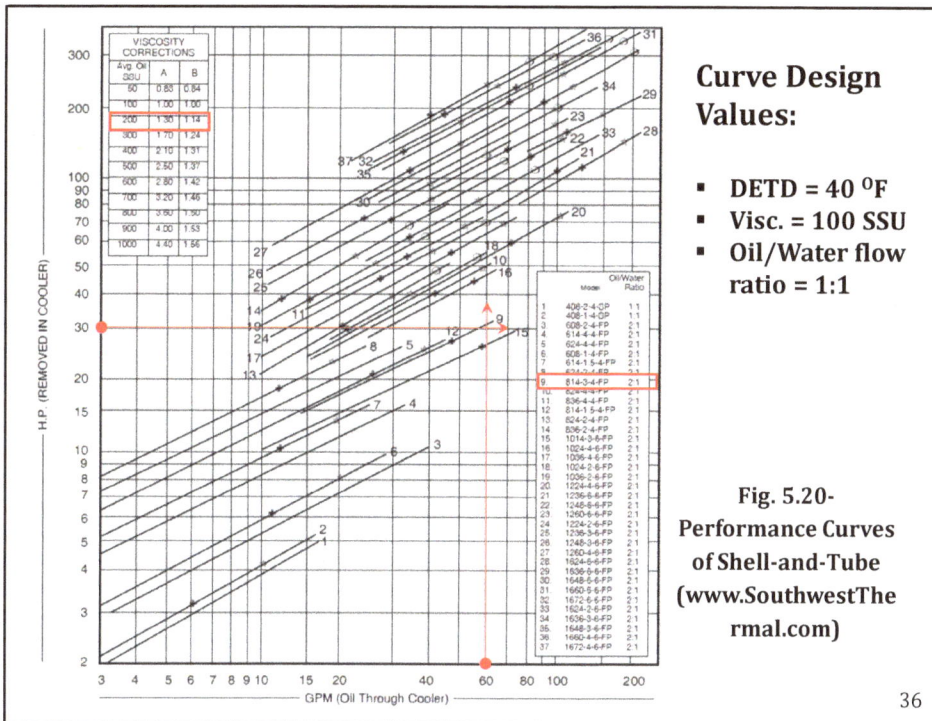

Curve Design Values:

- DETD = 40 $^\circ$F
- Visc. = 100 SSU
- Oil/Water flow ratio = 1:1

Fig. 5.20-
Performance Curves
of Shell-and-Tube
(www.SouthwestThe
rmal.com)

36

❖ **Step 1 (Calculation of the Design Cooling Capacity qd):**

- Column B (at Visc. = 200 SSU) →
- Cooling capacity correction factor, Kv = 1.14.

Equation given by manufacturer

Design Cooling Capacity q_d = qr x (DETD/EDT) x Kv   5.3

→ q_d = 19.8 x (40/30) x 1.14 = 30 hp

❖ **Step 2 (Selecting Proper Size Cooler):**

- Performance curves at q_d = 30 hp & **Qoil** = 60 gpm → → cooler size = 9

❖ **Step 3 (Check for the Actual Pressure Drop):**

- Design pressure drop (legend +) **DPoil** = 5 psi.
- Column A (at Visc. = 200 SSU) → ΔP correction factor, Kp = 1.3.

Equation given by manufacturer

Actual Pressure Drop Δp = DPoil x Kp   5.4

→ **Δp** = 5 x 1.3 = 6.5 psi < given allowable pressure drop (10 psi)

37

5.6- Plat-Type Oil Coolers

- Consist of a stack of stamped heat exchange plates, either →
  - Brazed together.
  - Bolted together in a frame with gaskets in between the plates.

5.6.1- Brazed Plate-Type Oil Coolers

❖ **Construction:**

- A stack of stamped stainless steel heat transfer plates.
- The plates are brazed with copper or nickel spacers.

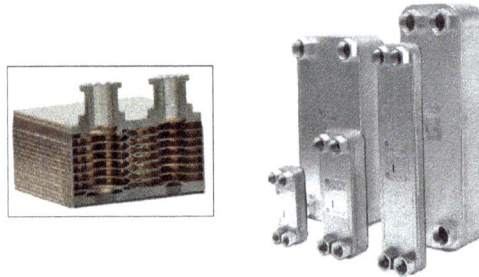

Fig. 5.21- Construction of Brazed Plate-Type Oil Cooler (Courtesy of Hydac)   38

38

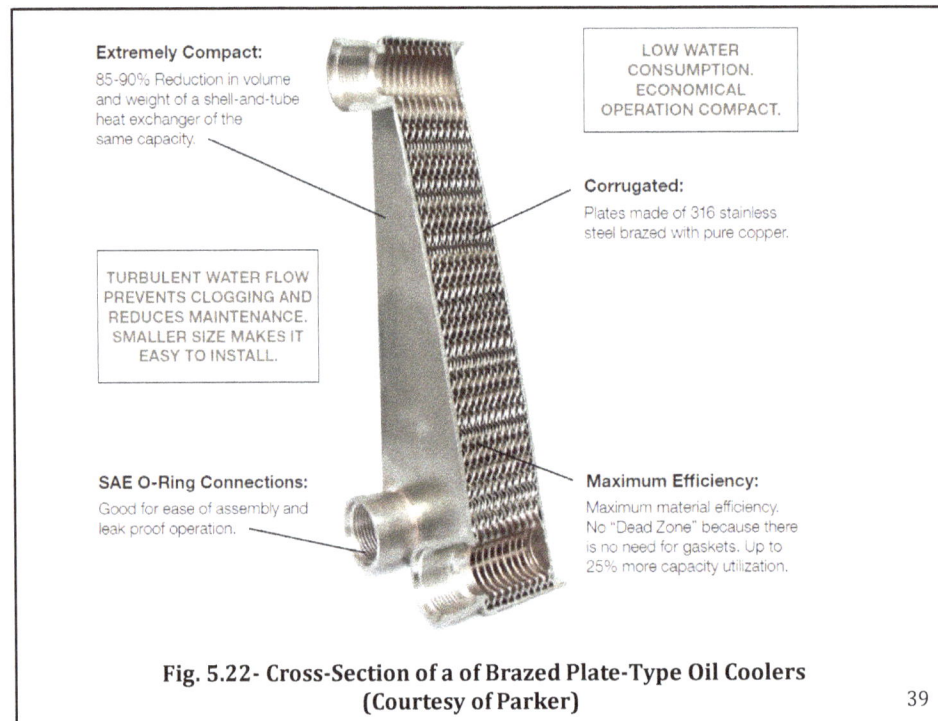

Extremely Compact:
85-90% Reduction in volume and weight of a shell-and-tube heat exchanger of the same capacity.

LOW WATER CONSUMPTION. ECONOMICAL OPERATION COMPACT.

TURBULENT WATER FLOW PREVENTS CLOGGING AND REDUCES MAINTENANCE. SMALLER SIZE MAKES IT EASY TO INSTALL.

Corrugated:
Plates made of 316 stainless steel brazed with pure copper.

SAE O-Ring Connections:
Good for ease of assembly and leak proof operation.

Maximum Efficiency:
Maximum material efficiency. No "Dead Zone" because there is no need for gaskets. Up to 25% more capacity utilization.

**Fig. 5.22- Cross-Section of a of Brazed Plate-Type Oil Coolers
(Courtesy of Parker)**   39

39

❖ **Operation:**
- Heat is transferred from the hot oil to the cold water through the plates.
- Special stamp pattern → turbulent flow →
- Optimum heat transfer and a self-cleaning.

❖ **Direction of Flow:**
- Concepts of parallel-flow and counter-flow are applicable.
- Inlets and outlets are clearly identified.
- Changing flow direction is easy and allowed.
- Better to have hot oil flowing through the outer plate.
- Otherwise → Water condensation on the outside surface →
- Possibility of water ingression into the system.

❖ **Features and Applications:**
- Compact, lightweight, efficient → Ideal for mobile applications.
- Limited fluid capacity, smaller heat transfer surface area →
- Less cooling capacity.

40

40

5.6.2- Gasketed Plate-Type Oil Coolers

❖ **Construction:**
1. A stack of corrugated steel (If sea water is used → titanium plates).
2. Gaskets for consecutive plates "A" and "B".
3. Front fixed end plate (cover).
4. Rear moveable end plate (cover). It moves backward for disassembling process.
5. Guide and carrying bars.
6. Fluid connections.
7. Clamping bolts.

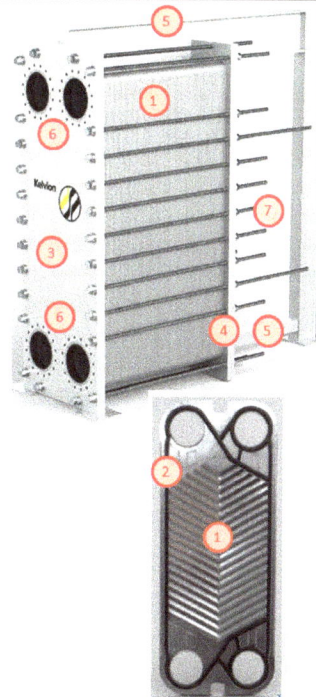

Fig. 5.23- Construction of Gasketed Plate-Type Heat Exchangers

41

41

❖ **Operation:**

Video 134 (11 min)

- Plates are marked for better identification.
- Plates are assembled in the correct order and so the gaskets →
- Perfectly aligned → when bolted →
- Gaskets are squeezed → clearance between plates are sealed →
- Flow channels are formed.

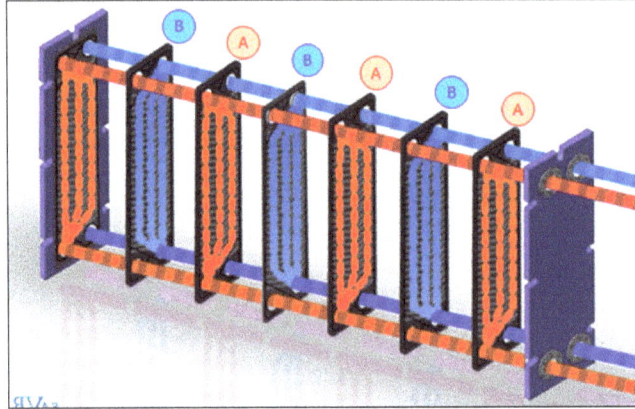

Fig. 5.24- Operation of Gasketed Plate-Type Heat Exchangers (savree.com) 42

42

- Number and size of the plates ↑→ cooling capacity ↑.
- # plates = odd number → # channels = even.
- Hot oil adjacent to the end plates →
  o Avoids condensing humidity on the end plates.
  o Promote heat dissipating from end plates to the surrounding.

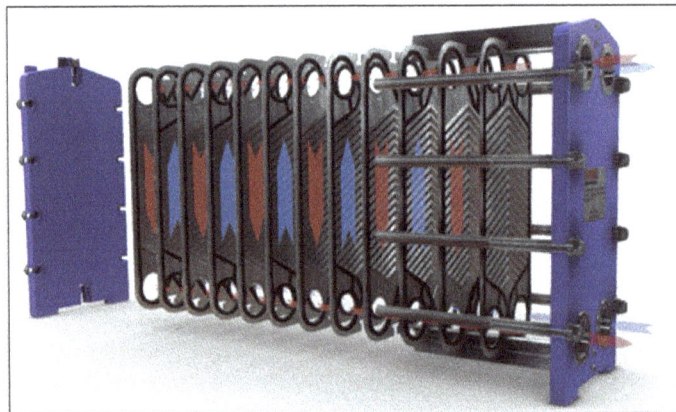

Fig. 5.24- Continue 43

43

❖ **Plate Design:**
▪ Plates corrugation pattern →
  ○ turbulent flow → improved heat transfer.
  ○ Strengthens the plate → thinner plates can be used.

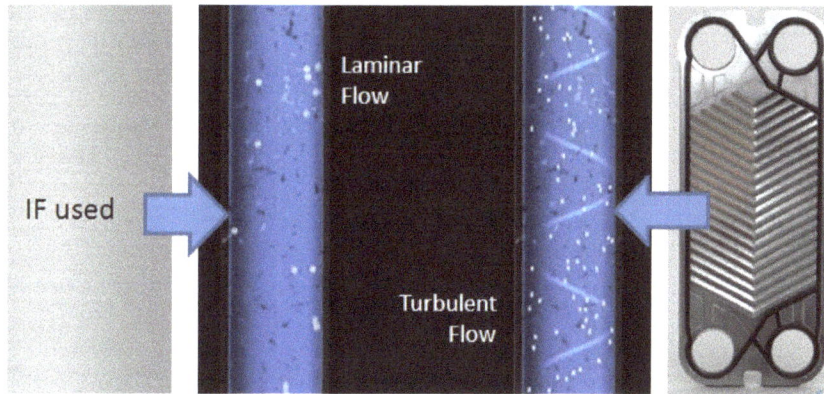

Fig. 5.25- Plate Design in Gasketed Plate-Type Heat Exchangers (savree.com)

44

44

❖ **Gasket Design:**
▪ Correct gasket → correct cooler operation.
▪ Leaking gaskets → heat transfer ↓ + mixing water with oil.

Video 437 (1.5 min)

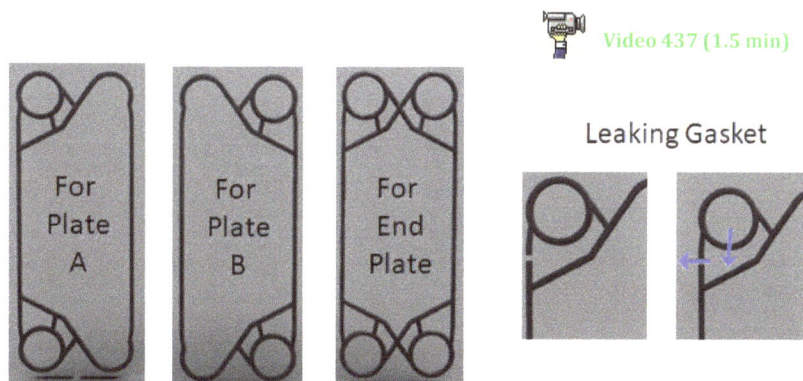

Fig. 5.26- Gasket Design in Gasketed Plate-Type Heat Exchangers (savree.com)

45

45

❖ **Flow Direction:**
- Parallel flow → nonhomogeneous thermal stress.
- Counter flow (ΔT ≈ constant) → better (cooling efficiency and reliability)
- Unlike brazed plat-type of coolers, complex plate stamp pattern →
- Changing flow direction isn't easy or allowed → consult manufacturer.

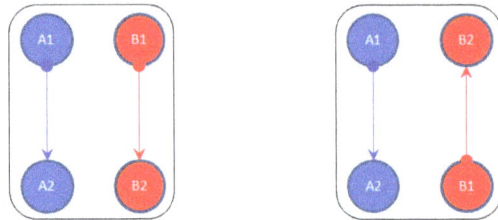

Fig. 5.27- Parallel-Flow versus Counter-Flow in Plate-Type Oil Coolers (Courtesy of Hydac)

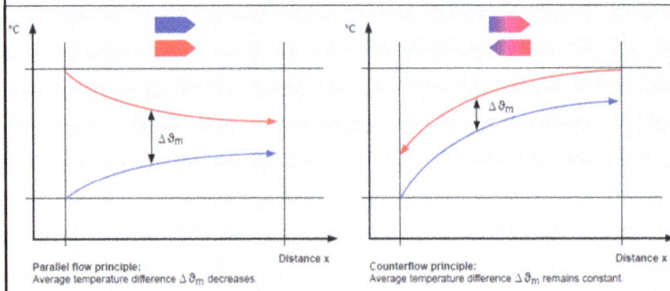

46

❖ **Heat Transfer:**
1. <u>Convective</u> heat transfer from the hot oil to the adjacent plates.
2. <u>Conductive</u> heat transfer through the wall of the plates.
3. <u>Convective</u> heat transfer from the plates to the cooling water.

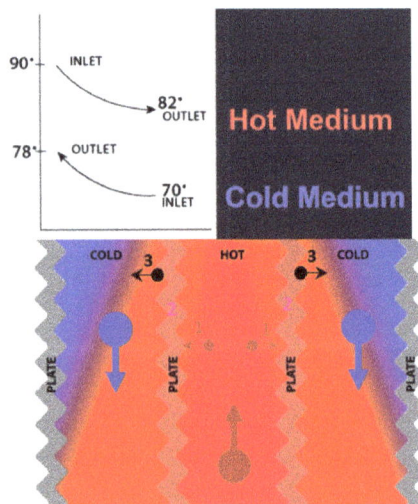

Fig. 5.28- Counter-Flow in Plate-Type Oil Coolers (savree.com)

47

❖ **Number of Passes:**

▪ the concepts of *single-pass* and *multi-pass* are applicable

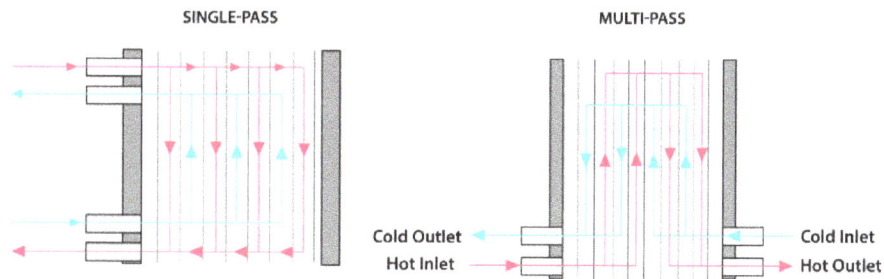

**Fig. 5.29- Single-Pass and Multi-Pass in Plate-Type Oil Coolers
(Courtesy of Hydac)**

48

48

❖ **Advantages of Gasketed Plate Heat Exchangers:**

▪ Available in wide range of sizes .

▪ Have high cooling capacities.

▪ Easy disassembly for cleaning and maintenance.

▪ Used with regular and sea water (titanium plates) .

❖ **Disadvantages of Gasketed Plate Heat Exchangers:**

▪ More expensive.

▪ Leak is difficult to locate.

▪ Replacing gaskets is problematic.

▪ Contamination → plate clearances are easily blocked by.

▪ Overtightening → gaskets oversqueezed → plate clearances ↓ → flow between the plates ↓→ (heat transfer ↓ + pressure drop↑).

49

49

5.6.3- Sizing of Plate-Type Oil Coolers

❖ **Performance Curves:**

- Developed by manufacturer under the following conditions:
  - Oil Viscosity = ISO VG 32.
  - Oil/Water Flow Ratio = 2:1.
  - Oil Inlet Temperature = 140 $^\circ$F.
  - Water Inlet Temperature = 80 $^\circ$F.
  - This means Design Entrance Temp Difference (DETD)
  - = 140 - 80 = 60 $^\circ$F.
- If a typical application has different conditions, correction must be made based on correction factors provided by the manufacturer.

Example:

❖ **Given Data:**

- Oil Viscosity is ISO VG 68.
- Oil Flow = 40 gpm.
- Available water flow 10 gpm.
- Required Cooling Capacity q_r = 40 hp (based on system energy analysis).
- Oil Inlet Temperature To = 140 °F.
- Water Inlet Temperature Tw = 100 °F.
- Allowable Pressure Drop = 30 psi.

50

❖ **Step 1 (Cooling Capacity and Pressure Drop Correction Factors):**

- Table 5.1 (provided by manufacturer) →
- Given oil viscosity is ISO VG 68 →
  - Cooling capacity correction factor (**Kv** = 1.2).
  - Pressure drop correction factor (**Kp** = 1.7).

Viscocity Class	Cooling Capacity Factor, Kv	Pressure Drop Factor, Kp
ISO VG 22	0.95	0.9
ISO VG 32	1.0	1.0
ISO VG 46	1.05	1.3
ISO VG 68	1.2	1.7
ISO VG 100	1.35	2.2
ISO VG 150	1.6	3.0
ISO VG 220	1.9	4.3

**Table 5.1- Cooling Capacity and Pressure Drop Correction Factors
(Courtesy of Parker)**

51

❖ **Step 2 (Entering Temperature Difference):**

- Actual Entrance Temperature Difference,
- **EDT** = To - Tw = 140°F - 100°F = 40°F

❖ **Step 3 (Entrance Temperature Difference Correction Factors):**

- Table 5.2 (provided by manufacturer) → **EDT** = 40°F → **Kt** = 1.43.

ETD	30	40	50	60	70
Kt	1.87	1.43	1.17	1.0	0.88

**Table 5.2- Temperature Difference Correction Factors
(Courtesy of Parker)**

52

❖ **Step 4 (Oil/Water Flow Ratio Correction Factors):**
- Given oil flow = 40 gpm & water flow = 10 gpm →
- Oil/water flow = 40/10 = 4 →
- Performance curve (Fig 5.30 check on smaller cooler size first) →
- Oil/water flow correction factor, **Kr** = 0.83.

❖ **Step 5 (Calculation of the Design Cooling Capacity q_d):**
   Design Cooling Capacity q_d = (qr x Kv x Kt)/Kr   5.5
→ q_d = (40 x 1.2 x 1.43)/0.83 = 83 hp

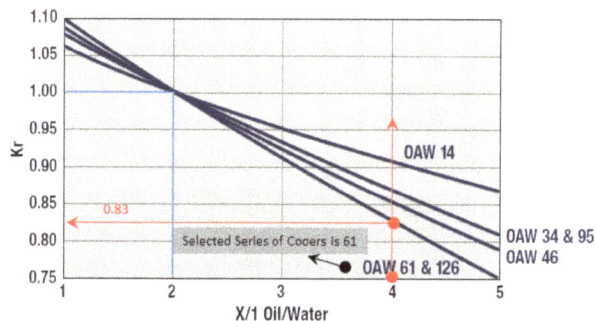

Fig. 5.30- Oil/Water Flow Correction Factors (Courtesy of Parker)   53

❖ **Step 6 (Selecting Proper Size Cooler based on Design Cooling Capacity q_d):**

- q_d = 83 hp & oil flow = 40 gpm → performance curves →
- Minimum size cooler is OAW 61-40.

OAW 46 & 61 **COOLING CAPACITY**

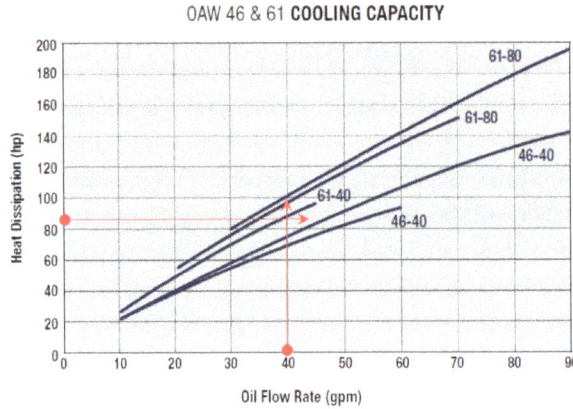

Fig. 5.31- Cooling Performance of Plate-Type Oil Coolers (Courtesy of Parker)

54

54

❖ **Step 7 (Check on Pressure Drop Δp based on Pressure Drop Correction Factor):**

- Performance curves (Fig. 5.32 provided by manufacturer) →
- Design pressure drop **DPoil** = 23 psi

OAW 46 & 61 **PRESSURE DROP**

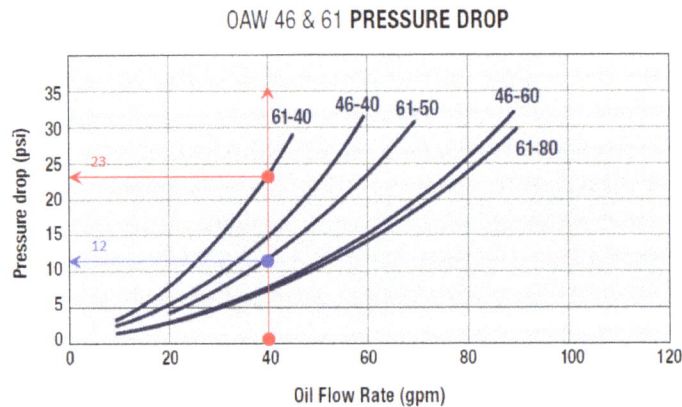

Fig. 5.32- Performance Curves of Plate-Type Oil Coolers (Courtesy of Parker)

55

55

Actual Pressure Drop $\Delta p = DPoil \times Kp$ **5.6**

- $\rightarrow \Delta p = 23 \times 1.7 = 39.1 >$ given allowable Δp (30 psi) $\rightarrow$
- Next larger size is (61-50) $\rightarrow$ **DPoil** = 12 psi $\rightarrow$
- Actual Pressure Drop $\Delta p = 12 \times 1.7 = 20.4 <$ given allowable Δp (30 psi)

56

56

5.7- Cooling-Filtration Units

- Plate-type cooler is integrated with an oil filter
- $\rightarrow$ Offline cooling-filtration unit.

Fig. 5.34- Cooling-Filtration Unit (Courtesy of Donalson)

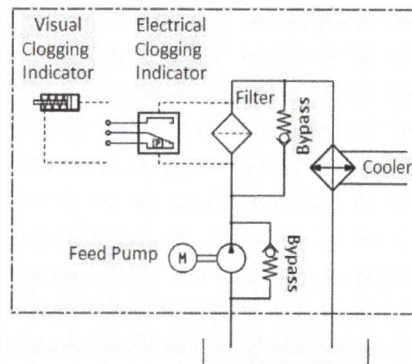

Fig. 5.33- Cooling-Filtration Unit (Courtesy of Hydac)

57

57

5.8- Oil Cooling Circuit Diagrams

- There is no single place for an oil cooler →
- It is a case-by-case type of solution.

- If a filter and a cooler are placed in series →
- The filter must b before the cooler →
  - Hot oil is filtered easier than cold oil.
  - Only clean oil entered the cooler.
  - Back pressure on cooler piping is less.

- Oil cooler protection (cracking pressure = 25 psi) by
  - Bypass valve.
  - Low-pressure non-adjustable relief valve.

58

58

5.8.1- Full Cooling on Return Line in Open Circuits
❖ **Features:**
- 100% of fluid volume is circulated through the cooler →
- Main pump is used.
- Cooler is sized based on the maximum flow in the circuit →
- Maximum flow could be > pump flow???

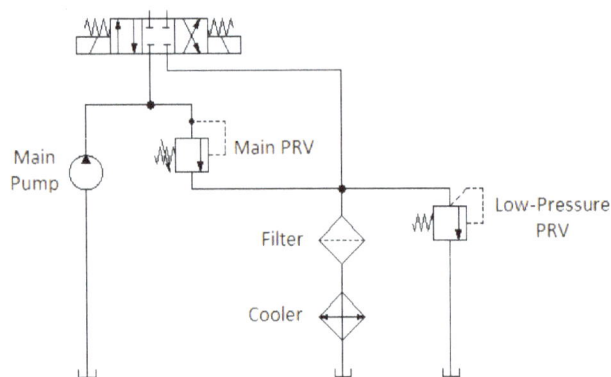

Fig. 5.35- Full Cooling on Return Line in Open Circuits

59

59

❖ **This solution is recommended when:**

- The main pump is small.

- The flow in the main circuit isn't surged by an accumulator or variable displacement pump.

❖ **The advantages of this solution are as follows:**

- Using the main pump →

- Saving the cost of additional setup for cooling and filtration.

- Fluid is circulated faster → heat is removed faster.

60

5.8.2- Full Offline Cooling

❖ **Features:**
- 100% of fluid volume is circulated through the cooler.
- Offline cooling circuit.
- Pump sizing rule of thumb →
- Full volume of oil in the tank is circulated every 10-15 minutes.
- Example: tank capacity = 150 liters → pump size =10-15 liter/min.
- Low pressure circuit → gear pump or even a centrifugal pump are used.

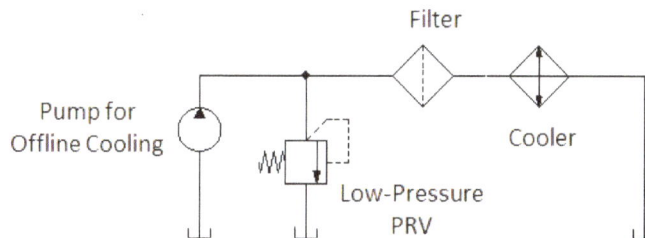

Fig. 5.36- Offline Full Cooling Circuit Diagram

61

❖ **This solution is recommended when:**

▪ The main pump is quite large.

▪ The flow in the main circuit isn't stable or is surged by an accumulator or variable displacement pump.

❖ **The advantages of this solution are as follows:**

▪ Less flow → smaller cooler size.

▪ A stable cooling and filtration performance irrespective of flow variation in the main circuit.

▪ The cooler is completely isolated from surge pressures in the return line that can potentially damage the cooler.

▪ Maintenance can be performed without the need to shut down the main system.

62

62

5.8.3- Partial Cooling on Boosting Pump Line in Open Circuits

❖ **Features:**

▪ A cooler is placed on the intake line.

▪ Part of oil volume passes through a cooler.

▪ Low-pressure relief valve is set at (5 to 10 psi) cracking pressure.

▪ **Concern:** introducing cold oil to the pump → concern of cavitation.

❖ **This solution is recommended when:**

▪ Main pump is large and requires intake boosting (supercharge).

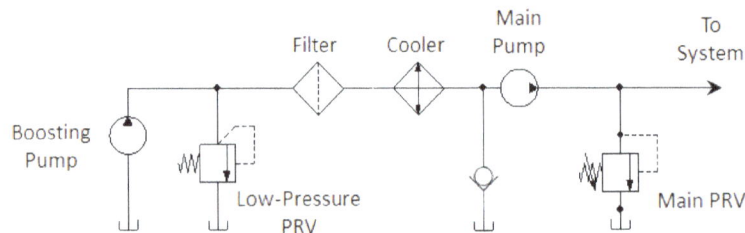

Fig. 5.37- Partial Cooling on Boosting Pump Line in Open Circuits 63

63

5.8.4- Partial Cooling on Main Relief Valve Return Line in Open Circuits

❖ **Features:**
- Cooler is placed on the return line after the main relief valve.
- Part of oil volume passes through a cooler.

❖ **This solution is recommended when:**
- Main RV is frequently opened → passes >= ¼ of pump flow.
- → main RV becomes the spot of the most heat generation.

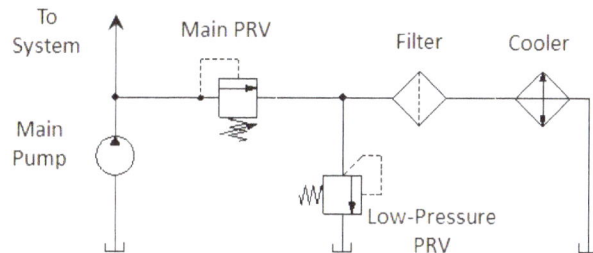

Fig. 5.38- Partial Cooling on Main Relief Valve Return Line in Open Circuits

64

64

5.8.5- Partial Cooling on Case Drain Line in Closed Circuits

❖ **Features:**
- Partial cooling.
- **Caution:** case drain back pressure.

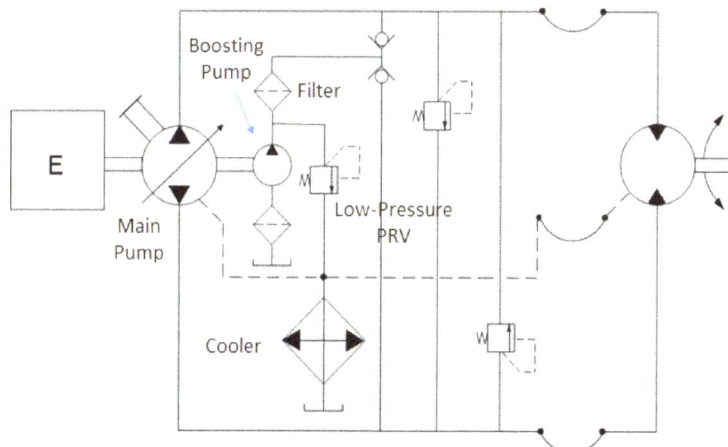

Fig. 5.39- Partial Cooling on Case Drain Line in Closed Circuits

65

65

5.9- Oil Temperature Automatic Control Solutions
5.9.1- ON/Off Automatic Oil Temperature Control

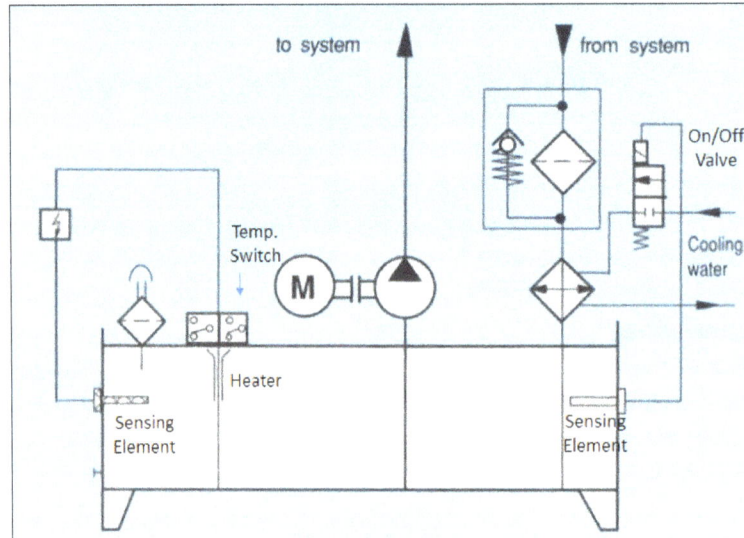

**Fig. 5.40- ON/Off Automatic Temperature Control Systems
(Courtesy of Bosch Rexroth)**

66

66

- **Cooling Water Control Valve:**
  o Operated by T Switch.

- **Cooling Water Shut-Off Valve** (not shown in the figure):
  o Shots cooling water for purposes of maintenance.

- **Temperature Sensing Elements:**
  o Should be placed in most representative point.
  o Should be placed inside a Sensing Bulb →
  o it can be removed for inspection without draining the reservoir.

- **Cooling Water Temperature Switch:**
  o T fluctuates in adjustable Threshold (Hysteresis) = Tmax - Tmin
  o T > Tmax → the switch energizes the solenoid valve → cooling water flows.
  o T < Tmin → the switch deenergizes the solenoid valve.

- **Air Cooling by Fan:**
  o T Switch → operates cooling fan OR coolant on/off valve.

67

- **Performance:**
  o Operating temperature fluctuates around a set point
  o Fluctuation range is proportional to the temperature switch threshold.

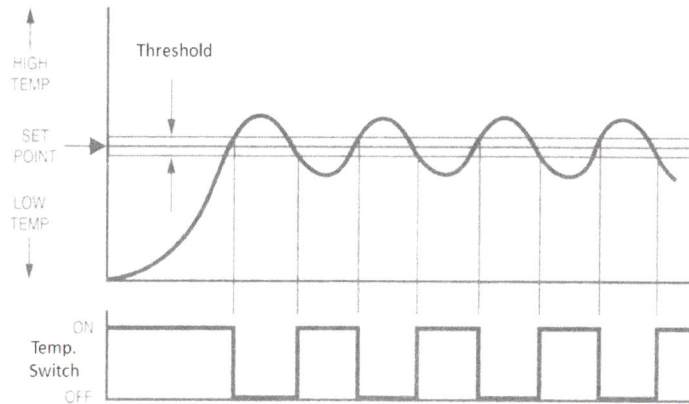

**Fig. 5.41- Performance of On/Off Automatic Temperature Control Systems
(Courtesy of Womack)** 68

68

5.9.2- Proportional Automatic Oil Temperature Control
- **Cooling Water Control Valve:**
  o Proportional valve → proportional control of cooling water flow.

- **Temperature Sensors:**
  o Sensor (analog output).

- **Performance:**
  o Closed Loop control (PID Controller) → Working T settles at the desired value with an acceptable SSE margin.

**Fig. 5.42- Performance of Proportional
Automatic Temperature Control Systems
(Courtesy of Womack)**

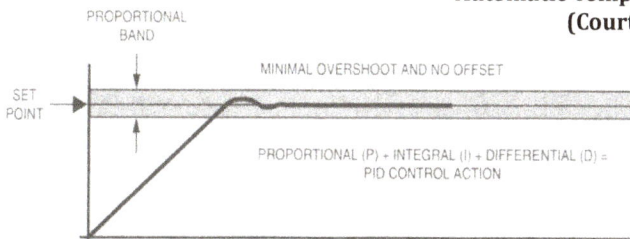

69

69

- **Air Cooling:**
  - o Electrical servo motor OR variable speed hydraulic motor
  - o → proportional fan speed control
  - o Alternatively, proportional valve → proportional control of coolant flow.

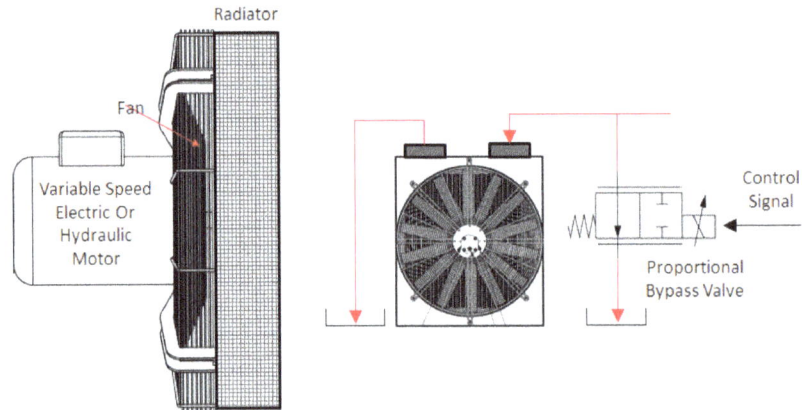

Fig. 5.42- Proportional Automatic Temperature Control Strategies for Air-Type Cooling Systems (Courtesy of Womack) 70

70

5.10- Electrical Oil Heaters
5.10.1- Construction and Operation of Electrical Oil Heaters

❖ **Heat Addition:**
- Passively (e.g. by running the pump over the relief valve).
- Actively by powering a heater immersed in the oil.

❖ **Construction:**
- Steel or stainless-steel enclosure "housing" →
- Heater removal without draining the reservoir.
- A built-in T switch or Sensor → On/Off or Proportional Control.

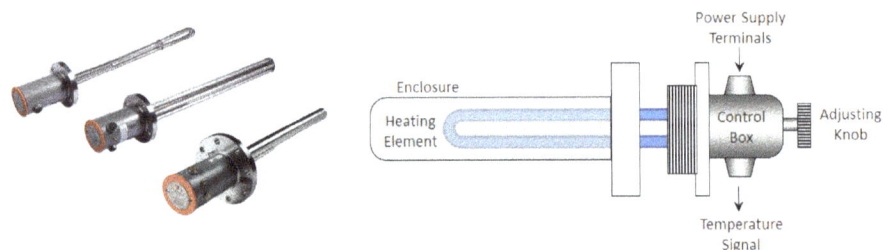

Fig. 5.44- Construction of Electrical Oil Heaters 71

71

❖ **Electrical Connections:**
- Choice of AC or DC power supply.
- Electrical terminals are sized proportional to the heater power.
- Electrical terminals are configured to meet the client's requirements

❖ **Operation:**
- The heater is immersed in the reservoir.
- Overheating petroleum-based hydraulic fluid → possible ignition
- → heater control system is interlocked with the pump
- → heat can't turn on unless there is oil flow through the reservoir.

❖ **Heater Placement:**
- Several small heaters is preferred over a one large size heater →
- Oil is heated homogeneously AND avoid local overheating in one spot.
- Should be located nearby sufficient oil flow → prevent overheating.

❖ **Heat Transfer in Electrical Oil Heaters:**
1. Convection: heating element → space inside the enclosure (bulb).
2. Conduction: through the walls of the enclosure.
3. Convection: walls of the enclosure → cold oil.

72

72

5.10.2- Sizing of Electrical Oil Heaters
❖ **Heating Capacity:**
- Design heating time (1-3 hours) → reasonable heater size.
- Knowing heating capacity → surface area.
- Surface area is limited to $1W/cm^2$ ($10\ W/in^2$) → avoid overheating.

HeatingCapacity(kW) =

$$\frac{\text{Tank Capcity (Gallons)} \times \left[\text{Desired min Temp.}(^0F) - \text{Ambient Temp.}(^0F)\right]}{800 \times \text{Design Heating Time (Hours)}}$$

5.7

❖ **Example:**
- Given:
  - Tank Capacity = 100-gallon
  - Ambient temperature = freezing temperature = 32 ^{0}F.
  - Minimum working temperature = 60 ^{0}F.
  - Design heating time = 1 hour.
- Eq. 5.3 → Heating Capacity (kW)
- = [100 (60-32)]/(800 x 1) = 3.5 x 1.341 = 4.7 kW.
- Heater Surface Area = 4700/2 = 2350 cm^2.

73

73

Chapter 5 Reviews

1. For a relief valve set to at 1714 PSI maximum pressure to protect a pump of 10 GPM, power wasted in the relief valve if it opens is?
    A. 10 kW.
    B. 10 HP.
    C. 10 Joule.
    D. 10 N.m.

2. In order to avoid local overheating at the place of the oil heater, the following rule of thumb is used to size the oil heater?
    A. Maximum heating capacity of 2 W/cm^2.
    B. Maximum heating capacity of 4 W/cm^2.
    C. Maximum heating capacity of 6 W/cm^2.
    D. Maximum heating capacity of 8 W/cm^2.

3. The best location for a filter-cooler assembly is?
    A. Filter located after the cooler.
    B. Cooler located before the filter.
    C. Filter located before the cooler.
    D. Does not make any difference.

4. Using proportional over on/off control mode is preferrable because?
    A. Temperature fluctuation will be less.
    B. Cooling water consumption will be less.
    C. Temperature sensor is used instead of temperature switch, so no need to set a differential range of temperature to turn the water valve ON and OFF.
    D. All the above.

5. The shown below oil cooler is considered a:
    A. Parallel flow oil cooler.
    B. Series flow oil cooler.
    C. Counter flow oil cooler.
    D. Cross flow oil cooler.

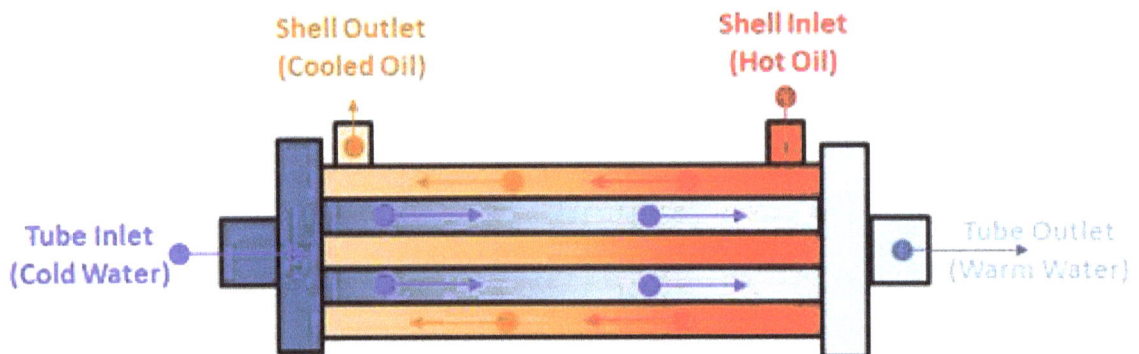

Chapter 5 Assignment

Student Name: --- Student ID: ------------------

Date: -- Score: -----------------------

Assignment:

A 200-gallon hydraulic reservoir is placed outdoor where the ambient temperature is around freezing temperature. The minimum temperature for such a machine is 60 °F. Allowable time for heating is one hour. Calculate the heating capacity and the heating surface area.

Chapter 6
Introduction to Filters

Objectives:

This chapter presents an overview of hydraulic filters including the contribution of filters in hydraulic systems, ISO1219 symbols, construction and operating principles. The chapter also presents various types of filters based on application in which the filter is used, type of connection to the circuit, body style of the filter, placement in the hydraulic circuit. The chapter also discusses the added accessories to the filter such as bypass valve and clogging indicators. Examples from industry are presented.

0

0

Brief Contents:

6.1 - Contribution of Filters in hydraulic Systems

6.2 - Types of Filters Based on Application

6.3 - Types of Filters Based on Types of Contamination

6.4 - Interpretation of ISO 1219 Symbols for Hydraulic Filters

6.5 - Basic Construction and Operation of Hydraulic Filters

6.6 - Types of Filters Based on Hydraulic Connections

6.7 - Types of Filters Based on the Filter Body Style

6.8 - Filter Clogging Indicators

6.9 - Types of Filters Based on their Placement in the Circuit

1

1

6.1 - Contribution of Filters in hydraulic Systems

- Fluid contamination → 80% of fluid systems failure.

- Hydraulic filters are classified based on:
  o Applications.
  o Place in a hydraulic circuit.
  o Body style.
  o Types of contamination.   Video 456 (0.5 min)

- Filters are available in various:
  o Sizes.
  o Dirt holding capacity.
  o Contamination removal efficiency.

2

2

- A properly selected filter must perform the following tasks:

  o Remove particulate contaminants (from the hydraulic fluid.
  o Maintain cleanliness level required by system manufacturer.
  o Remove chemical contaminants and their products (varnish, sludge, etc.).
  o Prevent aging of the hydraulic fluid due to chemical contaminants.
  o Maintain the lubricity of the fluid.
  o Extend the life of the hydraulic fluid.
  o Remove water content in the fluid.
  o Absorb moisture from breathing reservoirs.
  o Allow easy maintenance.
  o Increase component life and system reliability.
  o Increase intervals between scheduled maintenance.
  o Prevent unexpected failures and unplanned shutdown.

3

3

6.2 - Types of Filters Based on Application

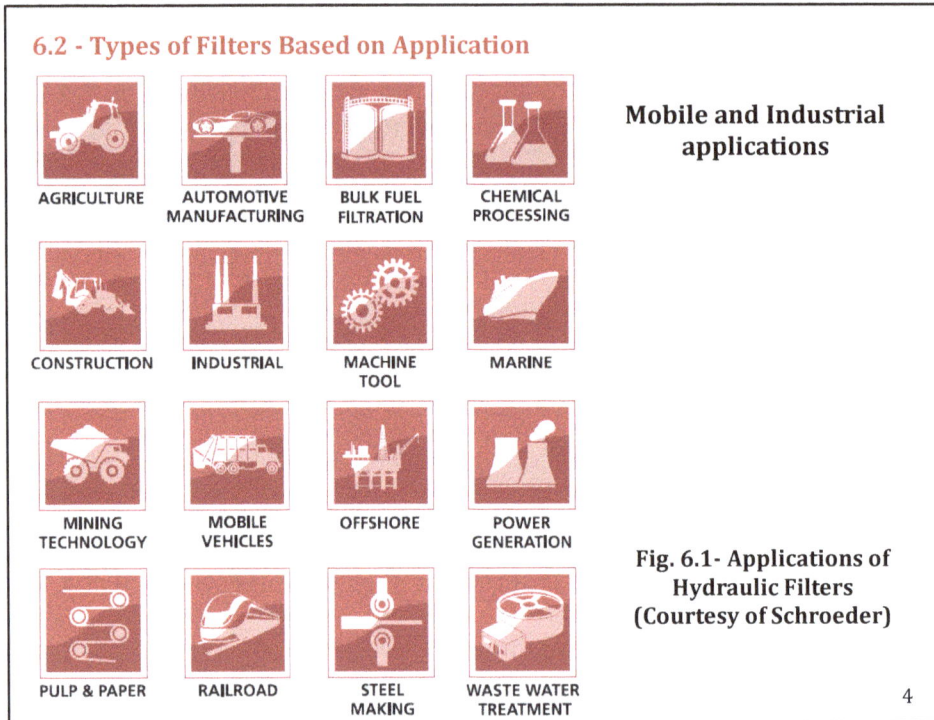

AGRICULTURE AUTOMOTIVE MANUFACTURING BULK FUEL FILTRATION CHEMICAL PROCESSING

CONSTRUCTION INDUSTRIAL MACHINE TOOL MARINE

MINING TECHNOLOGY MOBILE VEHICLES OFFSHORE POWER GENERATION

PULP & PAPER RAILROAD STEEL MAKING WASTE WATER TREATMENT

Mobile and Industrial applications

Fig. 6.1- Applications of Hydraulic Filters (Courtesy of Schroeder)

4

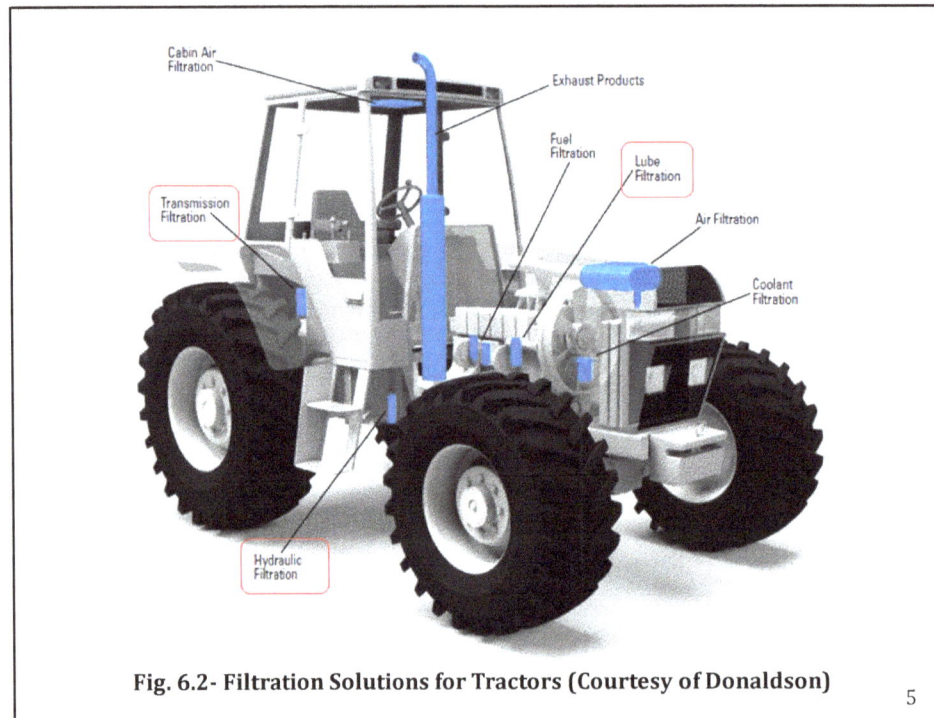

Cabin Air Filtration

Exhaust Products

Fuel Filtration

Lube Filtration

Air Filtration

Transmission Filtration

Coolant Filtration

Hydraulic Filtration

Fig. 6.2- Filtration Solutions for Tractors (Courtesy of Donaldson)

5

Video 433 (1 min)

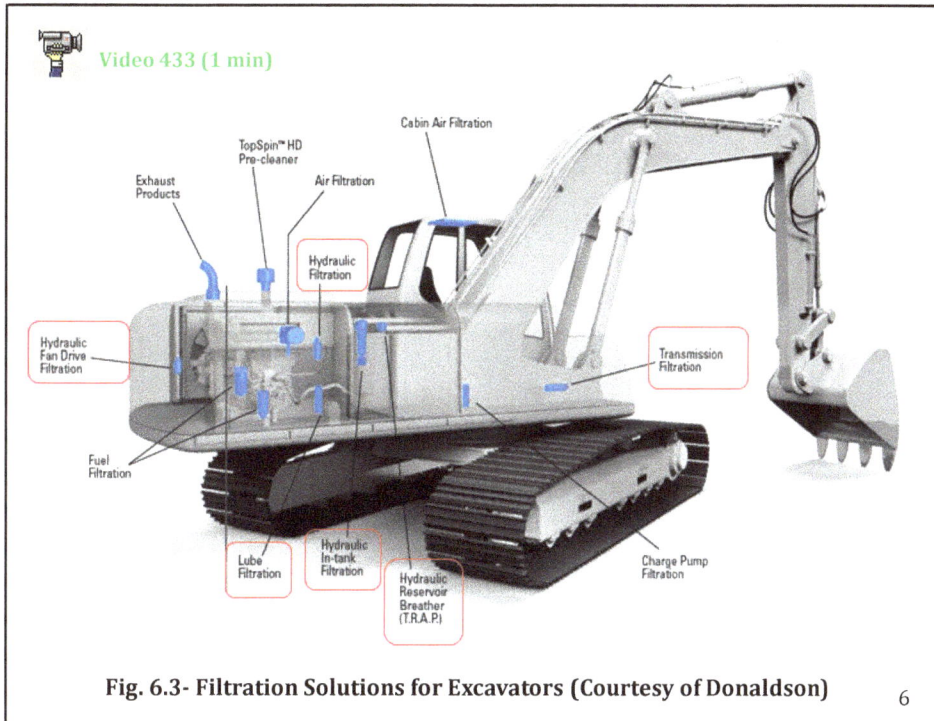

Fig. 6.3- Filtration Solutions for Excavators (Courtesy of Donaldson)

6

6

Video 449 (2 min)

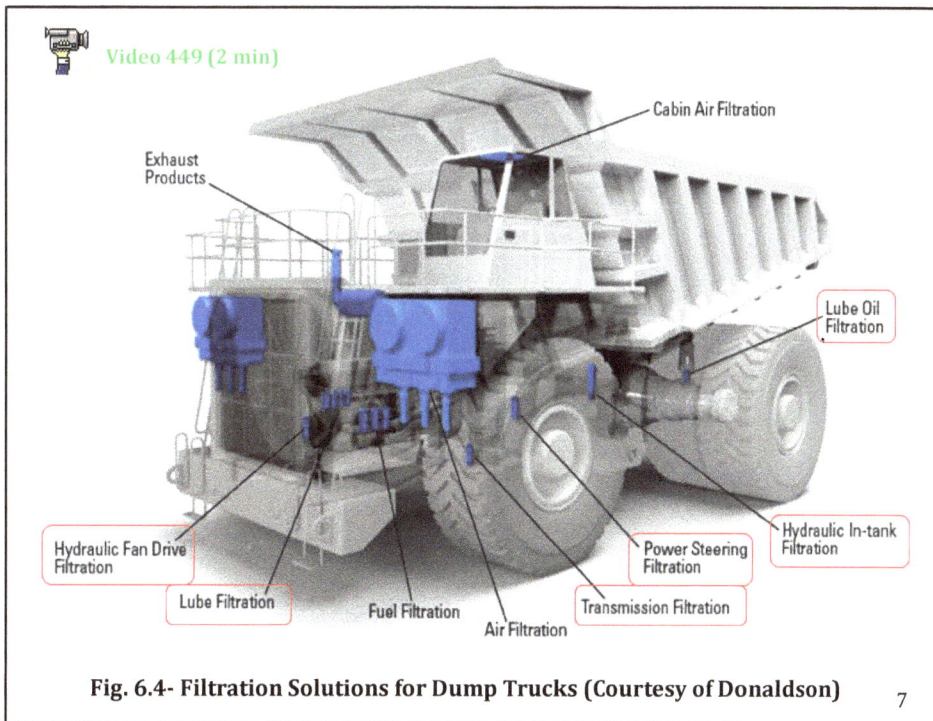

Fig. 6.4- Filtration Solutions for Dump Trucks (Courtesy of Donaldson)

7

7

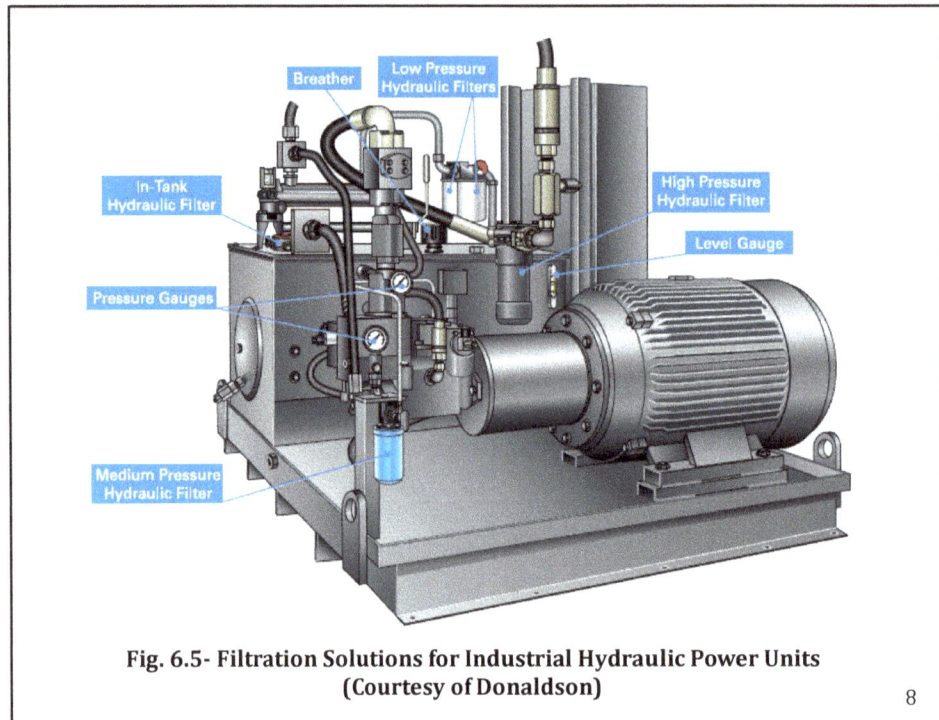

Fig. 6.5- Filtration Solutions for Industrial Hydraulic Power Units
(Courtesy of Donaldson)

8

8

6.3 - Types of Filters Based on Types of Contamination

Filters are classified based on the types of contamination as follows:

- Filters for <u>Fluidic Contaminants</u> → remove water content.

- Filters for <u>Chemical Contaminants</u> → remove fluid degradation products (oxidation, varnish and sludges).

- Filters for <u>Particulate Contaminants</u> → remove:
  - Solid or elastic.
  - Abrasive or nonabrasive.
  - Metallic or nonmetallic.

9

9

6.4 - Interpretation of ISO 1219 Symbols for Hydraulic Filters

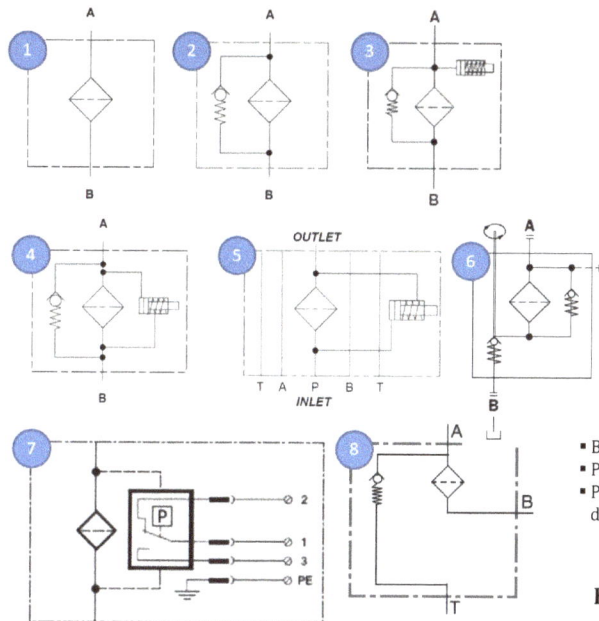

- **Self-Cleaning:**
  - o Manual or automatic mechanism.

  - o Used with surface type filter media.

  - o Wipes element outer surface and housing inner surface without stopping the machine.

- Bypass to tank →
- Protects element against collapse P.
- Protects sensitive components by directing dirty fluid to tank.

Fig. 6.6- Examples of Hydraulic Filters Symbols

10

10

6.5- Basic Construction and Operation of Hydraulic Filters

❖ **Filter Housing (Bowl):**
- Pressure vessel of the filter and
- Secures the element.
- Creates inlet/outlet seal.
- Has a drain plug for draining before disassembling the fitter.

❖ **Filter Head:**
- Filter mounting method.
- Contains other options:
  - o Bypass valve
  - o Clogging indicator.

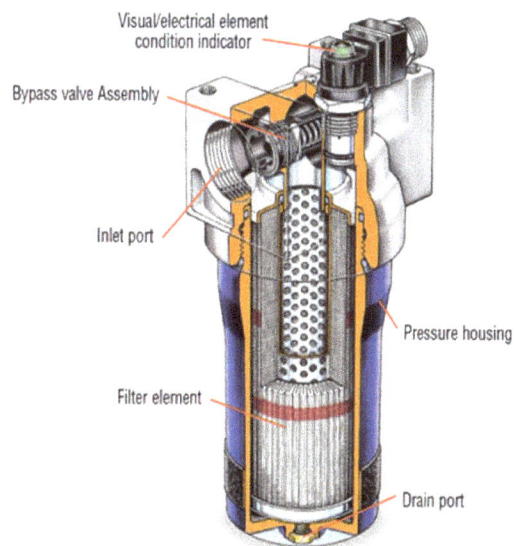

Fig. 6.7- Basic Construction of Hydraulic Filter (Courtesy of Parker)

11

❖ **Filter Element (Cartridge):**

▪ Central tube.

▪ Media (remove contaminants).

▪ End caps.

❖ **Clogging Indicator:**

▪ Alarming

▪ Control.

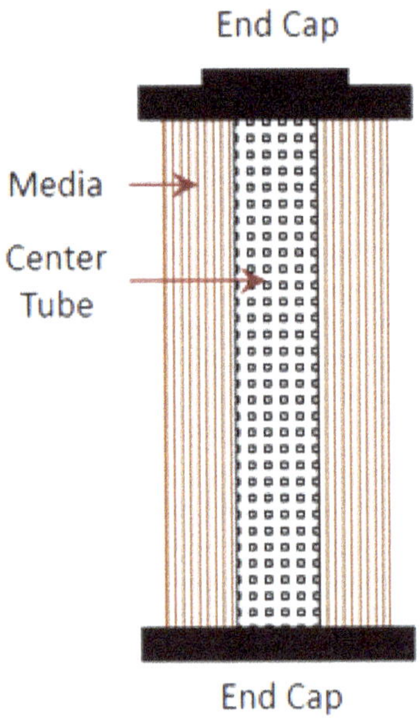

Fig. 6.7- Continue

12

❖ **Bypass Valve:**

▪ Bypass-Filters:

o Limits Δp across the element → prevents media collapse.

o Cracking pressure ≈ 1.5 – 7 bar (25 - 100 psi).

o Media clogged → bypass opened

o → all contaminants will get into the system.

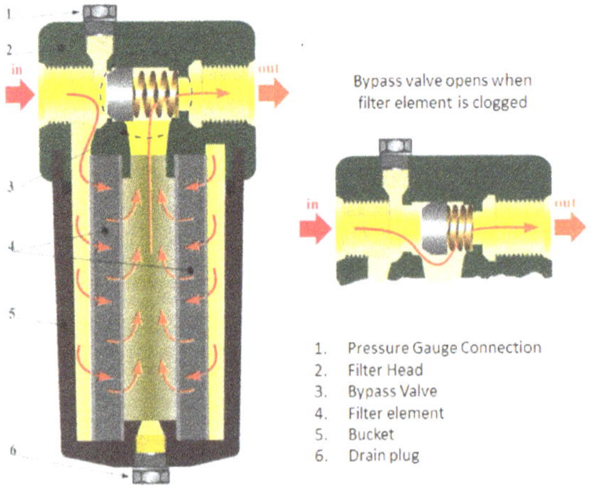

Bypass valve opens when filter element is clogged

Video 669 (1 min)

1. Pressure Gauge Connection
2. Filter Head
3. Bypass Valve
4. Filter element
5. Bucket
6. Drain plug

Fig. 6.8- Basic Construction of Hydraulic Filter (Courtesy of Assofluid)

13

- Non-Bypass Filters:
  o Sensitive components (servo and proportional valves) →
  o All flow must be filtered →
  o Filter element must have high collapse pressure.
  o Filter element must have Δp rating > max p on the filtration line.

- Bypass-to-Tank Filters:
  o Alternative to non-bypass filters.
  o Unfiltered bypass flow returns to tank through a third port.
  o Avoid using expensive high collapse pressure filter elements.

14

14

6.6 – Types of Filters Based on Hydraulic Connections

- Filter element can be replaced without disassembling the filter head from the system.

Line-Mounted   Flange-Mounted   Sandwich-Mounted

Fig. 6.9- Types of Filters Based on Hydraulic Connections

15

15

6.7 – Types of Filters Based on the Filter Body Style

- Various body styles → ease assembly + accessible during maintenance.

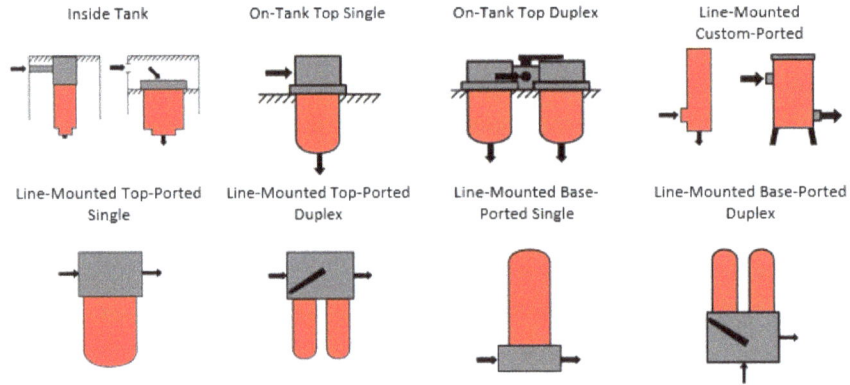

**Fig. 6.10- Types of Filters Based on the Filter Body Style
(Courtesy of Hydac)**

16

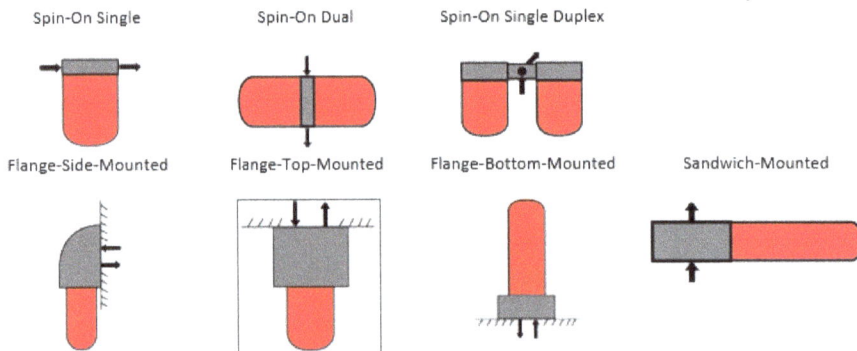

Fig. 6.10- Continue

17

6.7.1 – Inside Tank Filters

- The filter is completely installed inside the reservoir →
- Saves space, protects the filter, and reduces leak points.
- Concern: disassembling for maintenance.

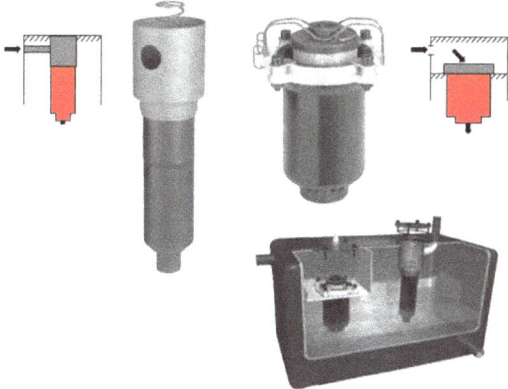

Fig. 6.11 Examples of Inside Tank Filters RFM-Series (Courtesy of Hydac)

18

18

6.7.2 – On-Tank Top Single Filters

- Filter head and the inlet/outlet ports are accessible from above the tank.
- Filter housing remains inside the tank → system design flexibility.

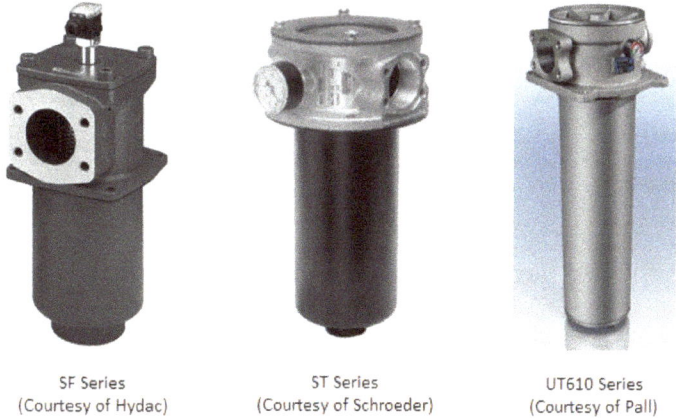

SF Series
(Courtesy of Hydac)

ST Series
(Courtesy of Schroeder)

UT610 Series
(Courtesy of Pall)

Fig. 6.12- Examples of On-Tank Top Single Filters

19

19

**Fig. 6.13- Example of On-Tank Top Single Filter - FIK Series
(Courtesy of Donaldson)**

20

6.7.3 – On-Tank Top Duplex Filters

- Both pressure and return filters are available in a duplex version.
- One filter works at a time → continuous, uninterrupted filtration.
- When servicing one side → duplex valve is shifted

www.behringersystems.com

www.hydraulicoilfilters.com

Fig. 6.14- Examples of On-Tank Top Duplex Filters

21

6.7.4 – Line-Mounted Top-Ported Single Filters

LF Series
(Courtesy of Hydac)

LMP900 - 901 Series
(Courtesy of MP Filtri)

CTF60 Series
(Courtesy of Schroeder)

Fig. 6.15- Examples of Line-Mounted Top-Ported Single Filters

22

22

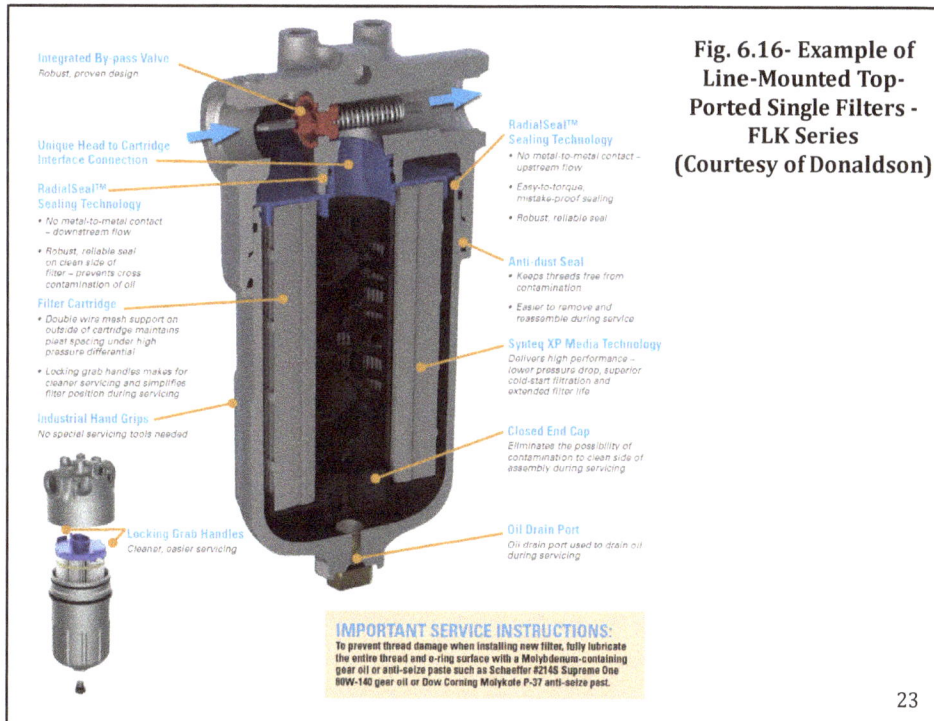

Integrated By-pass Valve
Robust, proven design

Unique Head to Cartridge
Interface Connection

RadialSeal™
Sealing Technology
• *No metal-to-metal contact – downstream flow.*
• *Robust, reliable seal on clean side of filter – prevents cross contamination of oil*

Filter Cartridge
• *Double wire mesh support on outside of cartridge maintains pleat spacing under high pressure differential*
• *Locking grab handles makes for cleaner servicing and simplifies filter position during servicing*

Industrial Hand Grips
No special servicing tools needed

Locking Grab Handles
Cleaner, easier servicing

RadialSeal™
Sealing Technology
• *No metal-to-metal contact - upstream flow*
• *Easy-to-torque, mistake-proof sealing*
• *Robust, reliable seal*

Anti-dust Seal
• *Keeps threads free from contamination*
• *Easier to remove and reassemble during service*

Synteq XP Media Technology
Delivers high performance – lower pressure drop, superior cold-start filtration and extended filter life

Closed End Cap
Eliminates the possibility of contamination to clean side of assembly during servicing

Oil Drain Port
Oil drain port used to drain oil during servicing

IMPORTANT SERVICE INSTRUCTIONS:
To prevent thread damage when installing new filter, fully lubricate the entire thread and o-ring surface with a Molybdenum-containing gear oil or anti-seize paste such as Schaeffer #214S Supreme One 80W-140 gear oil or Dow Corning Molykote P-37 anti-seize past.

Fig. 6.16- Example of Line-Mounted Top-Ported Single Filters - FLK Series (Courtesy of Donaldson)

23

23

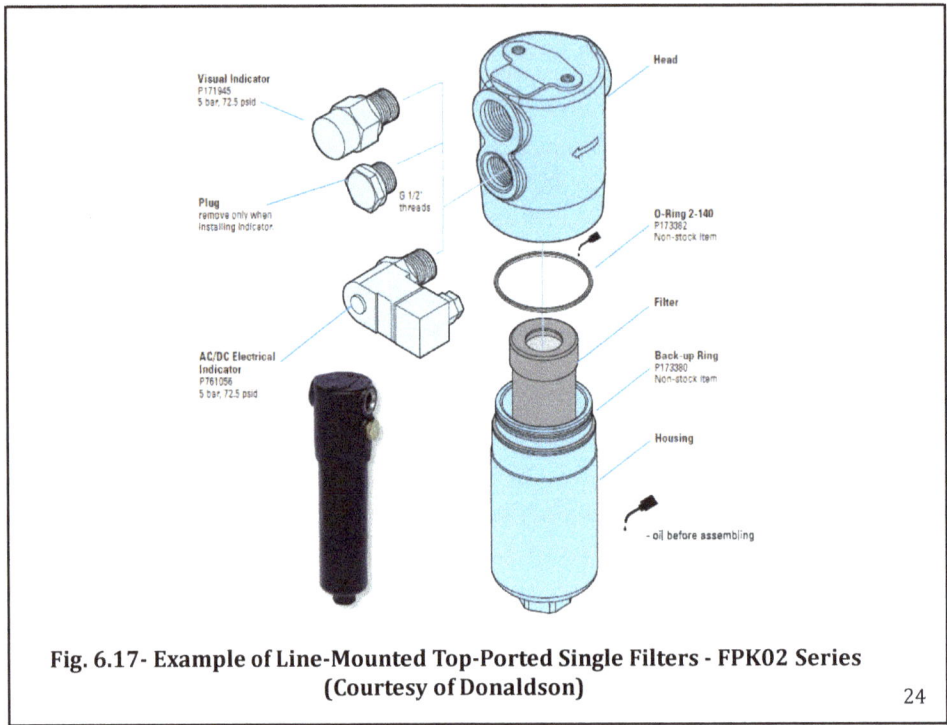

**Fig. 6.17- Example of Line-Mounted Top-Ported Single Filters - FPK02 Series
(Courtesy of Donaldson)**

24

24

6.7.5 – Line-Mounted Top-Ported Duplex Filters

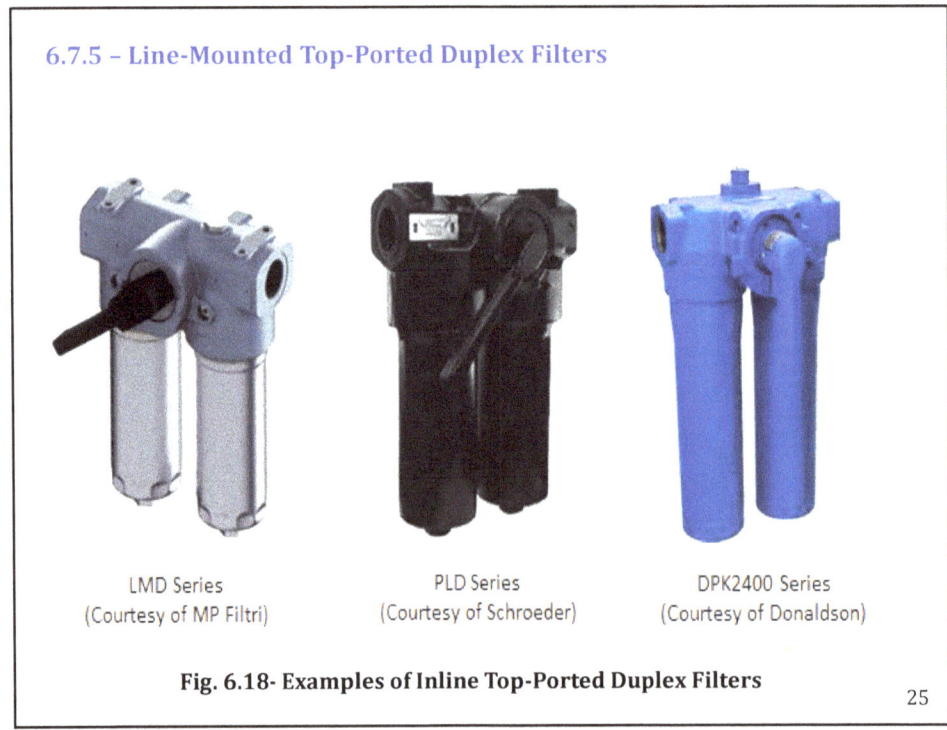

LMD Series
(Courtesy of MP Filtri)

PLD Series
(Courtesy of Schroeder)

DPK2400 Series
(Courtesy of Donaldson)

Fig. 6.18- Examples of Inline Top-Ported Duplex Filters

25

25

6.7.6 – Line-Mounted Base-Ported Single Filters

HF4P Series
(Courtesy of Hydac)

Series Athalon™ UH210
(Courtesy of Pall)

KF30 Series
(Courtesy of Schroeder)

Fig. 6.19- Examples of Line-Mounted Base-Ported Single Filters

26

26

6.7.7 – Line-Mounted Base-Ported Duplex Filters

Fig. 6.20- Example of Line-Mounted Base-Ported Duplex Filters - RFLDH Series (Courtesy of Hydac)

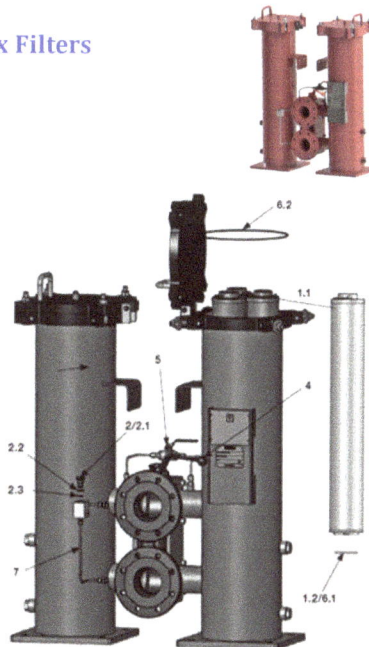

Item	Consists of	Designation
1.		Filter element
	1.1	Filter element
	1.2	O-ring
		No. of elements per filter side / size
2.		Indicator plug VD 0 A 1.0 /-V
	2.1	Clogging indicator or indicator plug
	2.2	Profile seal ring
	2.3	O-ring
3.		SEAL KIT VD/VM/VR/VR FKM
4.		Lever for change-over valve
5.		Equalization line ball valve
6.		SEAL KIT RFLD...FKM
	6.1	O-ring (element)
	6.2	Lid seal
7.		Indicator and equalization line pipe and plumbing

27

27

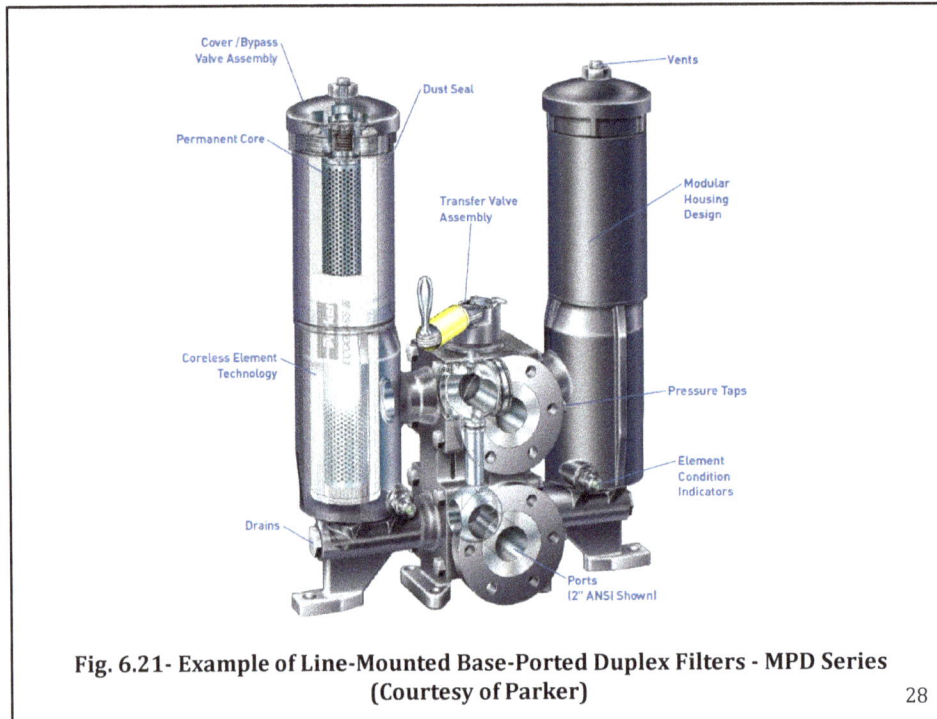

**Fig. 6.21- Example of Line-Mounted Base-Ported Duplex Filters - MPD Series
(Courtesy of Parker)**

28

28

6.7.8 – Line-Mounted Custom-Ported Filters

- The housing is made of rolled steel or stainless steel.
- ANSI flange connections for each filter size →
- Connection flexibility → no need for additional adapters.

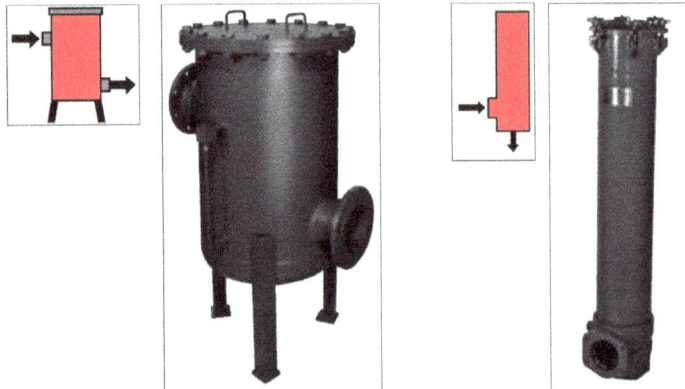

**Fig. 6.22- Examples of Line-Mounted Custom-Ported Filters – RFL Series
(Courtesy of Hydac)**

29

29

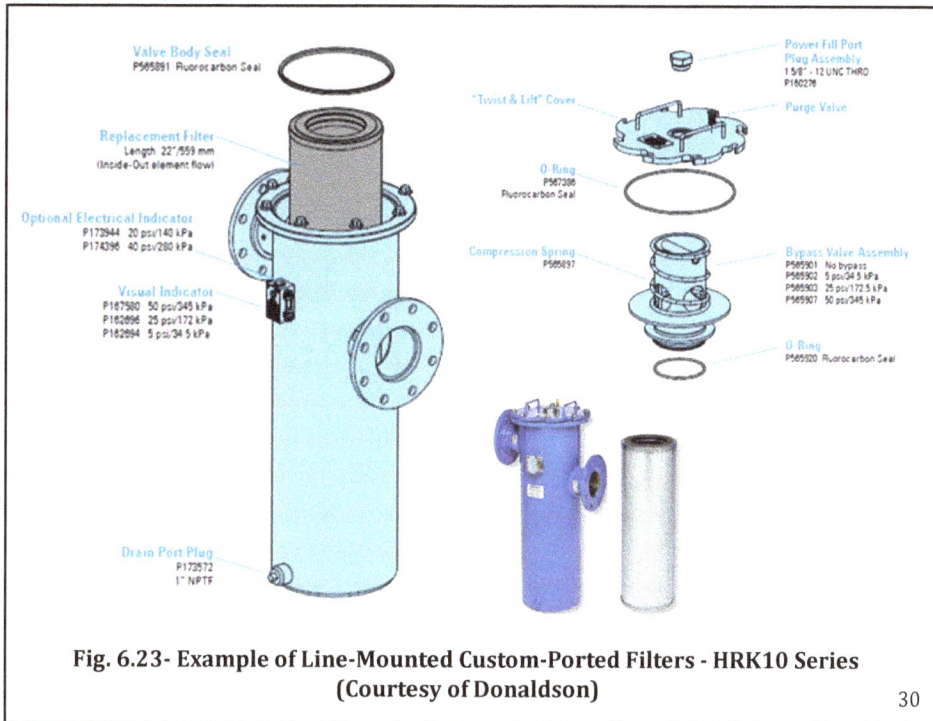

**Fig. 6.23- Example of Line-Mounted Custom-Ported Filters - HRK10 Series
(Courtesy of Donaldson)**

30

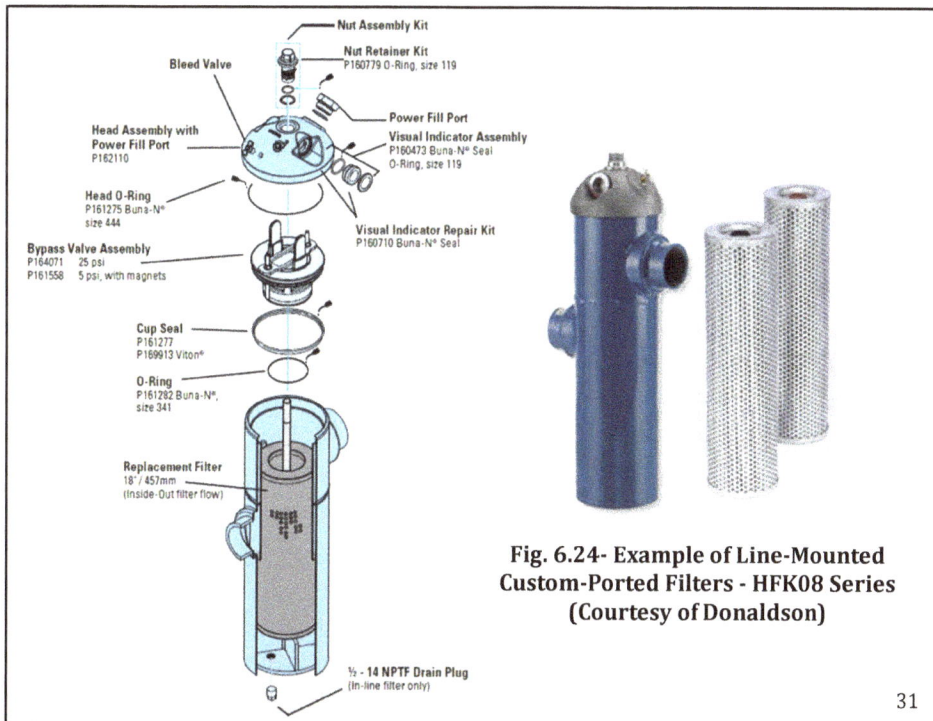

**Fig. 6.24- Example of Line-Mounted
Custom-Ported Filters - HFK08 Series
(Courtesy of Donaldson)**

31

6.7.9 – Spin-On Single Filters

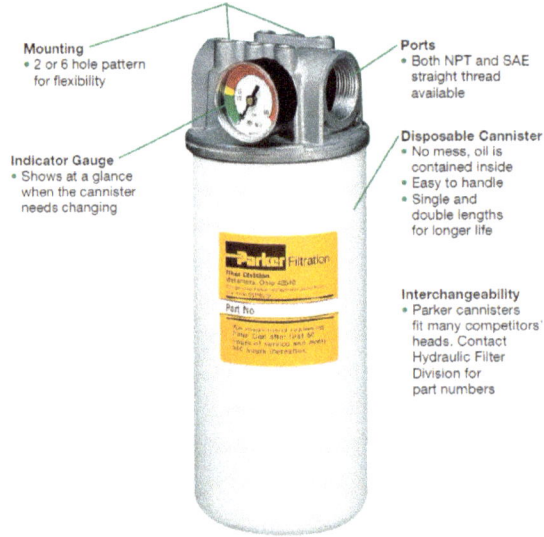

Mounting
• 2 or 6 hole pattern for flexibility

Indicator Gauge
• Shows at a glance when the cannister needs changing

Ports
• Both NPT and SAE straight thread available

Disposable Cannister
• No mess, oil is contained inside
• Easy to handle
• Single and double lengths for longer life

Interchangeability
• Parker cannisters fit many competitors' heads. Contact Hydraulic Filter Division for part numbers

**Fig. 6.25- Example of Spin-On Single Filters - 12AT/50AT Series
(Courtesy of Parker)**

32

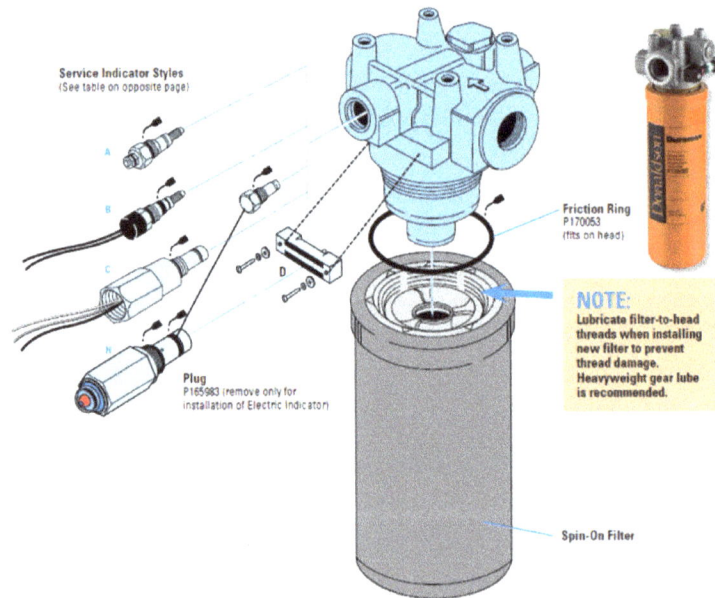

Service Indicator Styles
(See table on opposite page)

Plug
P165983 (remove only for installation of Electric Indicator)

Friction Ring
P170053 (fits on head)

NOTE:
Lubricate filter-to-head threads when installing new filter to prevent thread damage. Heavyweight gear lube is recommended.

Spin-On Filter

**Fig. 6.26- Example of Spin-On Single Filters - HMK03 Series
(Courtesy of Parker)**

33

6.7.10 – Spin-On Dual Vertical Filters

Dual → Dirt holding capacity ↑

SP100/120 Spin-On Filters

Working Pressures to:	150 psi 1035 kPa 10.3 bar	
Rated Static Burst to:	250 psi 1725 kPa 17.2 bar	
Flow Range to:	100 gpm 379 lpm	

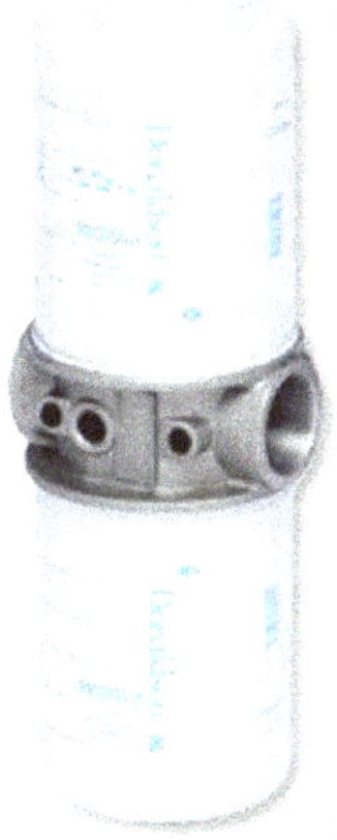

Fig. 6.27- Example of Spin-On Dual Vertical Filters
(Courtesy of Donaldson)

34

34

6.7.11 – Spin-On Dual Horizontal Filters

Video 409 (0.5 min)

SP80/90 Spin-On Filters

Working Pressures to:	150 psi 1035 kPa 10.3 bar	
Rated Static Burst to:	250 psi 1725 kPa 17.2 bar	
Flow Range to:	100 gpm 379 lpm	

Fig. 6.28- Example of Spin-On Dual Horizontal Filters
(Courtesy of Donaldson)

35

35

6.7.12 – Flange-Side-Mounted Filters

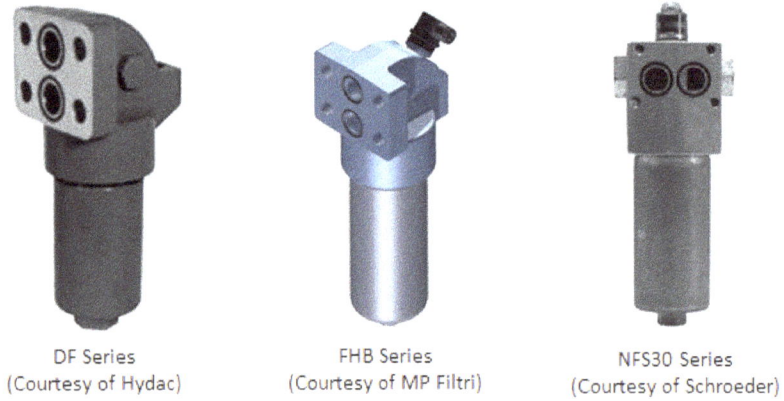

DF Series
(Courtesy of Hydac)

FHB Series
(Courtesy of MP Filtri)

NFS30 Series
(Courtesy of Schroeder)

Fig. 6.29- Examples of Flange-Side-Mounted Filters

36

36

6.7.13 – Flange-Top-Mounted Filters

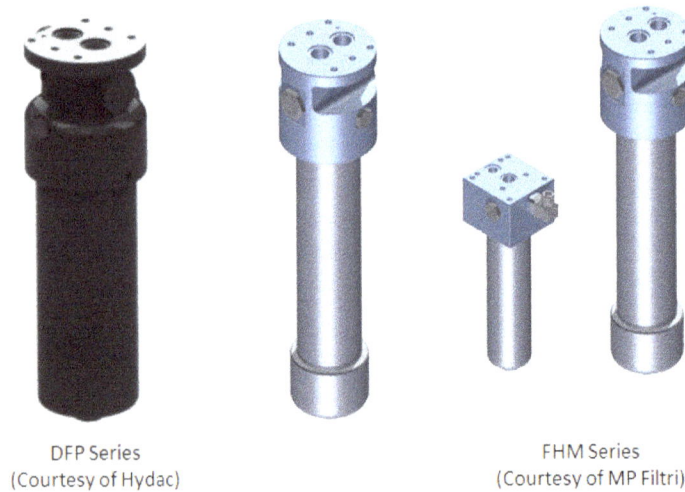

DFP Series
(Courtesy of Hydac)

FHM Series
(Courtesy of MP Filtri)

Fig. 6.30- Examples of Flange-Top-Mounted Filters

37

37

6.7.14 – Flange-Bottom-Mounted Filters

Cover
- Handle protects indicators from damage
- Easy on, easy off, for fast service

Air Bleed
- Helps protect bearings and other sensitive components from trapped air

Fill Port
- Prefilter the fluid, before it gets into the machine's system
- Purge air while filling

Indicators
- You can tell element condition at a glance
- Both visual and electrical available

Bowl
- Rugged cold drawn steel — excellent fatigue resistance
- Three sizes for any application: Single (8"), Double (16"), and Triple (39")

Ports
- SAE straight thread or flange face

Bypass Valve (not visible)
- Soft seat design for zero internal leakage
- Located in cover assembly

Drain Port (not visible)
- Clean and easy servicing
- Lets you drain bowl of fluid before element changes

Fig. 6.31- Examples of Flange-Bottom-Mounted Filters (Courtesy of Parker)

38

6.7.15 – Sandwich-Mounted Filters

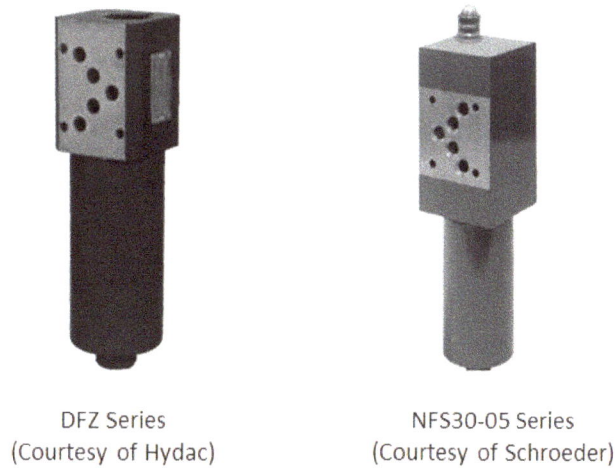

DFZ Series
(Courtesy of Hydac)

NFS30-05 Series
(Courtesy of Schroeder)

Fig. 6.32- Examples of Sandwich-Mounted Filters

39

6.7.16 – Screw-In Filters

- Also called *Manifold-Mounted* or *Cartridge-Style* filters.
- installed in special cavities to protect critical components.
- They are not intended to replace main filters.

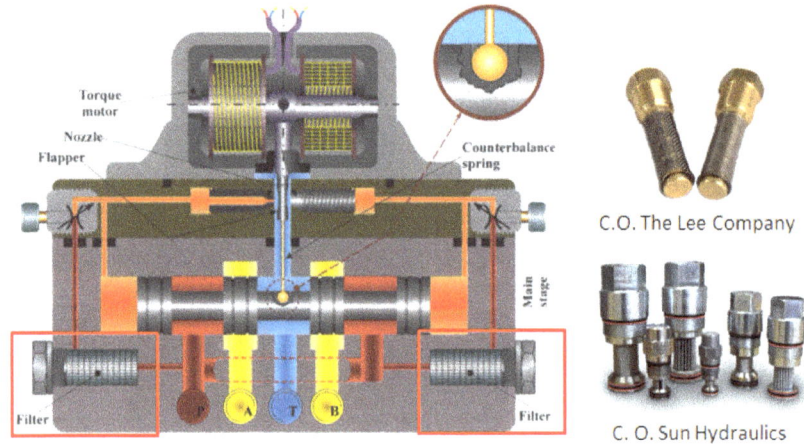

C.O. The Lee Company

C. O. Sun Hydraulics

**Fig. 6.33- Example of Screw-In Filters – CP-C16 Series
(Courtesy of Hydeck)**

40

40

6.8 - Filter Clogging Indicators

❖ **Purpose of Clogging Indicators:**

- Rate of dirt ingression isn't same for all machines →
- Changing filters based on (fixed time OR # of hours of operation) →
  - ○ Filter may be replaced while it is relatively clean → Not cost effective.
  - ○ Filter may be replaced while it is relatively dirty → Risky.
- Clogging indicators → visual or electrical warning devices.

❖ **Advantages of using Clogging Indicators:**

- Eliminates guessing of changing the element.
- Avoids the unnecessary cost of replacing elements too soon.
- Protect sensitive components that is intolerant to contamination.

41

41

❖ **Configurations of Clogging Indicators:**

**Fig. 6.34- Various Configurations of Clogging Indicators
(Courtesy of MP filtri)**

42

42

❖ **Differential vs. Static Pressure Visual Indicator:**

▪ **Δp Indicators:**

○ Flow through the filter → **Δp** (housing + element) > **Δpmax** → activated.

○ Used in most pressure and inline filters.

▪ **Static Pressure Indicators:**

○ Flow through the filter → Static **p** at upstream of the filter → activated.

○ Used for filters where downstream **p** is ambient.

○ (return and offline filters).

○ **Note:** if components are located downstream of the filter → false reading.

43

43

❖ **Clogging Indicator Settings:**
- The indicator is set to trip before:
  o The element becomes fully clogged.
  o Pressure = bypass cracking pressure →
- The operator sufficient time to take corrective action.

- Examples from Hydac literature:
  o Bypass Filter → trip at 15 psi (1 bar) < bypass $P_{cracking}$.
  o Non-bypass Filter → trip at 15 psid (1bard) < element changeout Δp.

- → A HYDAC Pressure Filter:
  o Bypass valve begins to crack at 87 psid (6 bard) →
  o Indicator is set to trip at 72 psid (5 bard).

- →A HYDAC Return Filter:
  o Bypass valve begins to crack at 43 psi (3 bar) →
  o Indicator is set to trip at 29 psi (2 bar).

44

44

❖ **Manual vs. Automatic Reset:**

- Electrical Clogging Indictors:
o After changing filter element → Automatic reset.

- Visual Clogging Indicator:
o After changing filter element → Automatic reset.
o If Manual reset → System shutdown → indicator still shows that the filter is dirty.

45

❖ **Interchangeability:**
- Various types of clogging indicators can be used without the need for special mechanical arrangement.

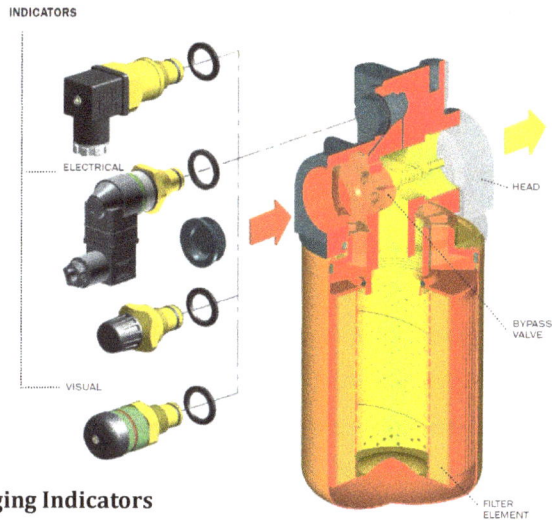

Fig. 6.35- Examples of Clogging Indicators (Courtesy of Assofluid)

46

6.8.1- Visual Clogging Indicators

❖ **Operation of Visual Δp Clogging Indicators:**
- **Δp** across the filter > trip **Δp** →
- Piston/magnet assembly is driven down against a spring →
- Attractive force between the magnet and indicator pin ↓
- Indicator pin rises → Visual indication that the filter must be serviced.
- **Δp** across the filter < trip **Δp** → automatic reset.

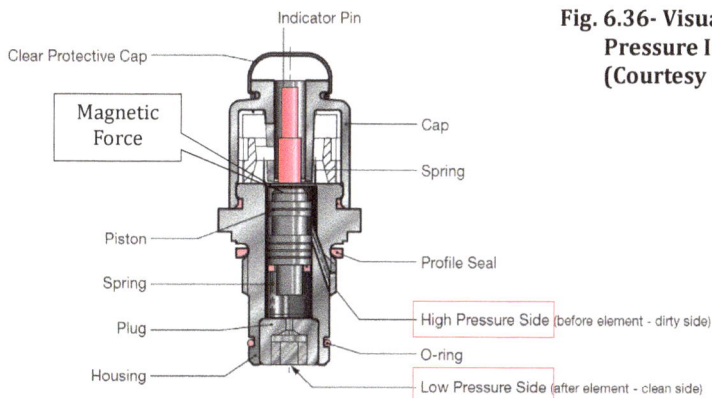

Fig. 6.36- Visual Differential Pressure Indicators (Courtesy of Hydac)

47

❖ **Operation of Static Pressure Visual Indicators:**
▪ Pressure upstream of the filter acts upon a diaphragm →
▪ Pressure > trip pressure →
▪ Indicator pin rises against the spring → Visual indication
▪ Δp across the filter < trip Δp → automatic reset.

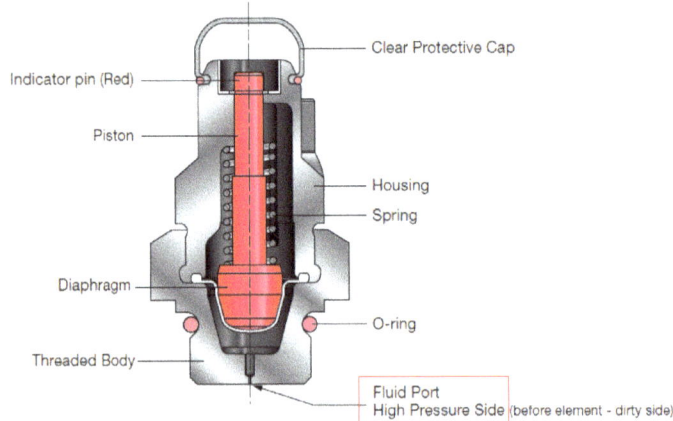

Fig. 6.37- Static Pressure Visual Indicators (Courtesy of Hydac)   48

48

6.8.2- Electrical Clogging Indicators

❖ **Concepts:**
▪ Electrical signal → actuate an electric switch →
  o Alarming: Warning light or buzzing sound.
  o Control: Stop the machine to protect sensitive components.

❖ **Thermal Lockout:**
▪ Machine cold start → fluid viscosity is high → Δp is high →
▪ False clogging signal →
▪ Electric indicators are equipped with thermal lockout.
▪ Prevents indicator tripping when T < specified T.
▪ How it Works: (AND Function).
▪ Bi-metal strip **T** switch in series in the circuit.

❖ **Single Pole, Double Throw Switches (SPDT):**
▪ Electrical indicators contain single-pole, double-throw switches →
▪ Choice of normally open (NO) or normally closed (NC) contacts.

49

49

❖ **Operation of Differential Pressure Electrical Clogging Indicator:**
- Δp across the filter ↑→
- The piston is driven down against a spring.
- Tripping causes a switch to make or break →
- Warning or Control signal for servicing.

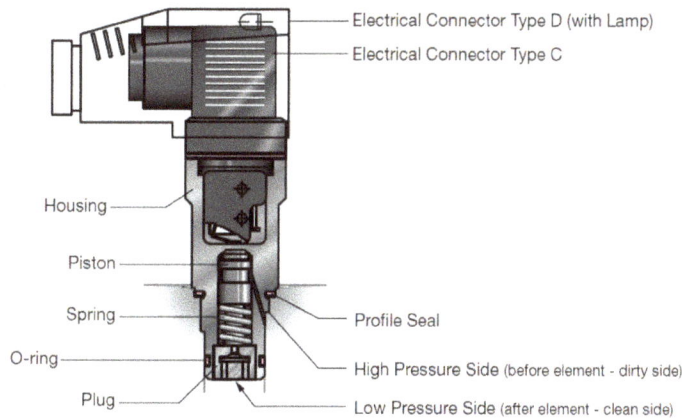

Fig. 6.38- Differential Pressure Electrical Indicators
(Courtesy of Hydac)

50

50

6.9 – Types of Filters Based on their Placement in the Circuit

- **Online Filters:** are placed in series (1 through 4)
- **Offline Filters:** are placed in parallel (5 & 6)

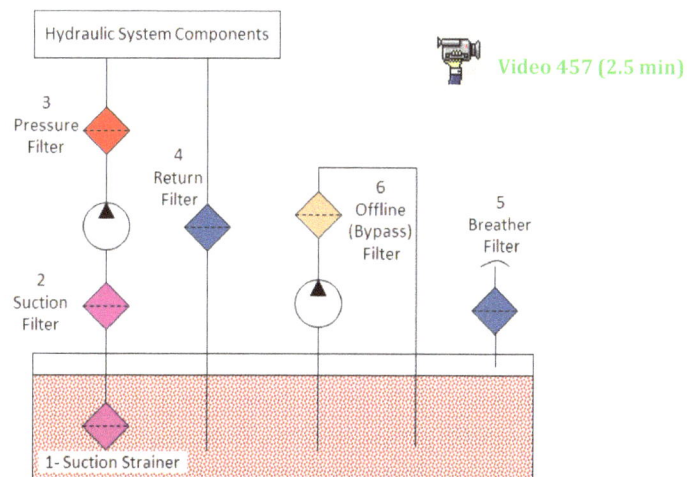

Video 457 (2.5 min)

Fig. 6.39- Types of Filters Based on their Placement in the Circuits

51

51

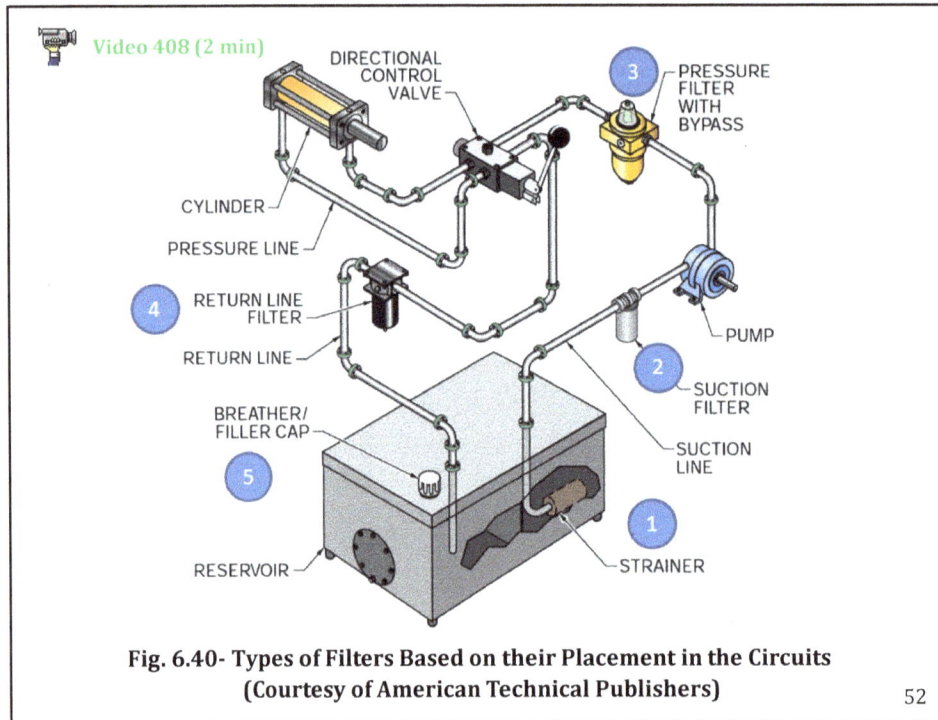

**Fig. 6.40- Types of Filters Based on their Placement in the Circuits
(Courtesy of American Technical Publishers)**

52

52

6.9.1- Suction Strainers

❖ **Placement:**

- Are connected to the beginning of the suction line before a suction filter
- Aligned with the suction line or a 90° elbow.

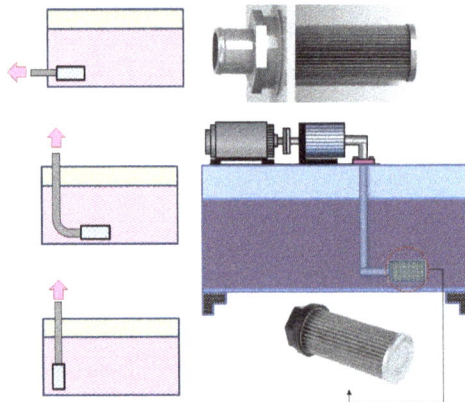

Fig. 6.41 – Suction Strainers

53

53

❖ **Primary Duty:**
- Capture the relatively large particles before getting into the pump.
- Because of cavitation concerns →
  o Not recommended for large flow pumps
  o Approval from the pump manufacturer is required.

❖ **Cost:**
- Not a complex housing → least expensive among filters.

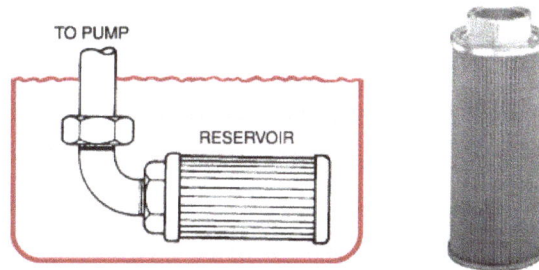

Fig. 6.42 – Suction Strainers (Courtesy of Schroeder)

54

54

❖ **Construction:**
- Optimized pleat size and screen area → extended life and low **Δp**.
- Some could be magnetic → capture metallic contaminants or wear products.

One pieced high strength nylon hex cap for reduced cost.

Cap assembly epoxy bonded to body for superior strength.

Pleated stainless wire cloth provides excellent flow with minimum pressure drop. Choice of different mesh sizes.

Sides of end caps are reverse tapered to enhance epoxy bonding.

Inner perforated steel support tube for added strength and rigidity.

Optional 3 or 5 PSI relief valve to prevent failure should the screen become clogged with debris.

Fig. 6.43 – Suction Strainers (ohfab.com)

55

55

❖ **Micron Size vs. Mesh Size:**

- Micron Size (Pore Size) →

- largest particles (μm) that can pass through the screen.

- Mesh Size (Porosity) → # of holes in one squared inch →

- Mesh Size ↑→ Micron Size ↓ (i.e. finer filter)
- If no mesh size is reported by the pump manufacturer →
- 250-500 mesh size is recommended.

❖ **Flow:** a strainer receives the full flow of the main pump flow.

56

56

❖ **Surface Area:**
- A strainer is sized based on the pump flow →
- Minimize Δp & avoid cavitation →
- Review the pump data sheet if found.
- Consult the pump manufacturer.
- If no information is found → Rule of thumb
- Surface area > 2 square inches for every GPM of the pump flow
- ≈ 3 cm² for every liters/min of the pump flow

Suction Strainer Surface Area (cm²) = 3 x Qp (lit/min) 1.1A

Suction Strainer Surface Area (in²) = 2 x Qp (gpm)   1.1B

- **Example:**
  o A 20 GPM pump.
  o Eq. 1.1B → A minimum strainer surface area of 40 square inches.

57

57

288/335

❖ **Magnetic Suction Strainers:**
- Magnetic suction strainers → dual protection.
- Capture large particles + attract ferrous particles of all sizes.

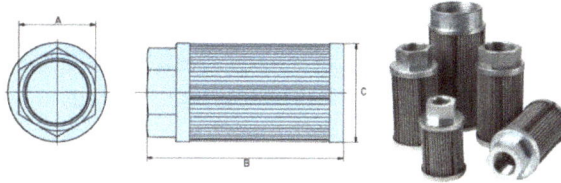

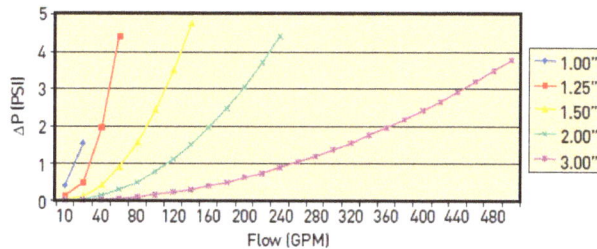

Fig. 6.44 – Magnetic Suction Strainers (Courtesy of Parker)

58

58

6.9.2- Suction Filters

❖ **Placement:**
- Before the inlet port of the pump.
- Between the pump and a suction strainer (if found)
- Mounted externally (outside the reservoir).

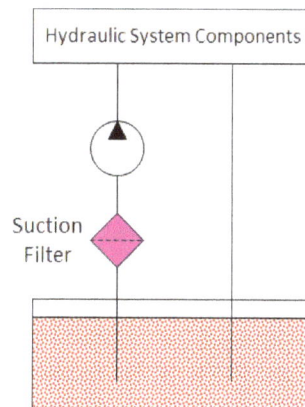

Fig. 6.45 – Placement of Suction Filters

❖ **Primary Duty:**
- Protect the pump.
- Concerns of cavitation → not recommended for large flow pumps → subject to manufacturer approval.
- If no information → maximum acceptable Δp <= 0.1 bar (1.45 psi).

59

59

❖ **Cost:** strainers < suction filter < pressure filters.

❖ **Construction:** Typically (housing, a filter element, and a filter head).

❖ **Bypass Valve:**
- Is highly recommended in suction filters → avoiding cavitation.
- Expensive pumps → vacuum switch to stop the machine.

❖ **Mesh Size:** A suction filter is coarse (60 to 250 micron) → limit Δp.

❖ **Flow:** Full flow of the main pump → sized based on the main pump flow.

60

60

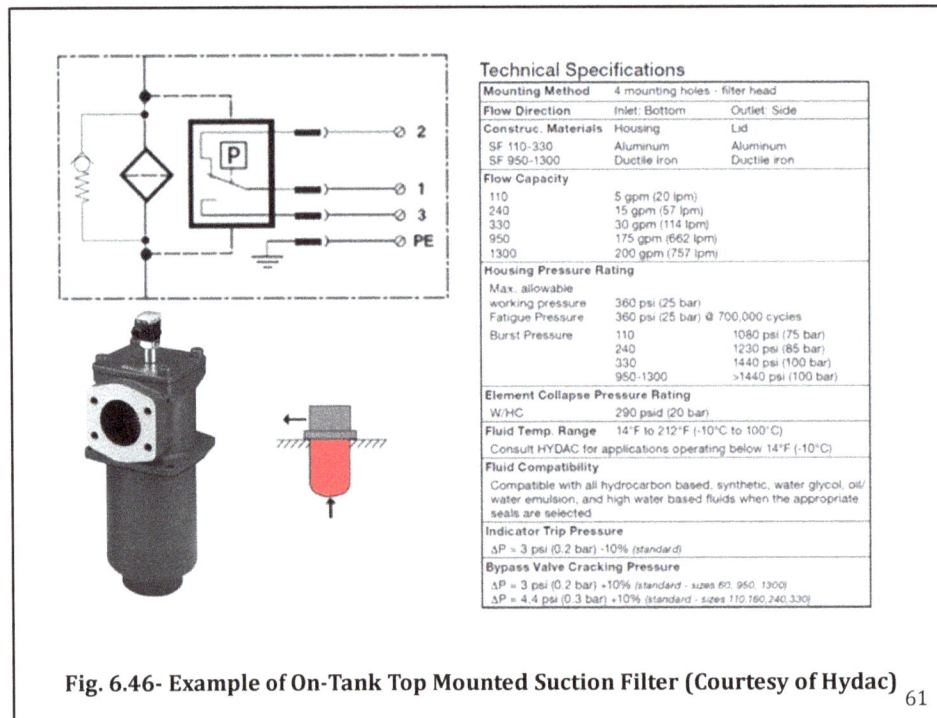

Technical Specifications

Mounting Method	4 mounting holes - filter head	
Flow Direction	Inlet: Bottom	Outlet: Side
Construc. Materials	Housing	Lid
SF 110-330	Aluminum	Aluminum
SF 950-1300	Ductile iron	Ductile iron
Flow Capacity		
110	5 gpm (20 lpm)	
240	15 gpm (57 lpm)	
330	30 gpm (114 lpm)	
950	175 gpm (662 lpm)	
1300	200 gpm (757 lpm)	
Housing Pressure Rating		
Max. allowable working pressure	360 psi (25 bar)	
Fatigue Pressure	360 psi (25 bar) @ 700,000 cycles	
Burst Pressure	110	1080 psi (75 bar)
	240	1230 psi (85 bar)
	330	1440 psi (100 bar)
	950-1300	>1440 psi (100 bar)
Element Collapse Pressure Rating		
W/HC	290 psid (20 bar)	
Fluid Temp. Range	14°F to 212°F (-10°C to 100°C)	
Consult HYDAC for applications operating below 14°F (-10°C)		
Fluid Compatibility		
Compatible with all hydrocarbon based, synthetic, water glycol, oil/water emulsion, and high water based fluids when the appropriate seals are selected		
Indicator Trip Pressure		
ΔP = 3 psi (0.2 bar) -10% (standard)		
Bypass Valve Cracking Pressure		
ΔP = 3 psi (0.2 bar) +10% (standard - sizes 60, 950, 1300)		
ΔP = 4.4 psi (0.3 bar) +10% (standard - sizes 110,160,240,330)		

Fig. 6.46- Example of On-Tank Top Mounted Suction Filter (Courtesy of Hydac)

61

61

6.9.3- Pressure Filters

❖ **Placement:**

- <u>Open Circuits</u> → Unidirectional Pressure Filters downstream the pump.
- <u>Closed Circuits</u> → Bidirectional Pressure Filters →
- Check valve rectifier built into the filter head → handle reserve flow.

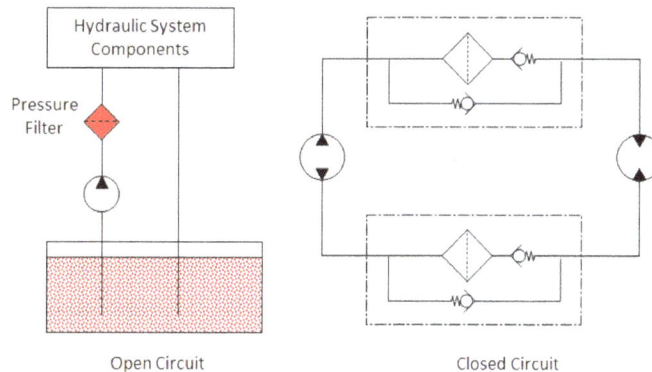

Fig. 6.47 – Placement of Pressure Filters

❖ **Primary Duty:**
- Captures pump wear products before it is spread →
- Protect the sensitive components in the system (prop. and servo valves).

❖ **Cost:**
- Special housing (sealing and carrying high **p**) → most expensive.

❖ **Mesh Size:** Based on required cleanliness class required.

❖ **Flow:** full flow of the main pump.

❖ **Bypass Valve:**
- Optional.
- Sensitive components → Non-Bypass filters →
- If media collapse P < maximum working pressure → Bypass-to-Tank filters could be used → dirt didn't go to system.

❖ **Construction:**

▪ Same components as suction and return filters.

▪ Housing must withstand (Pmax + P spikes + P fluctuation).

▪ Steady state pressure ratings are 100 - 400 bar (1500 psi to 6000 psi).

1. High strength ductile iron filter head with integral indicator port

2. Steel bowl with standard drain port

3. Proprietary element endcap assembly includes bypass and reverse flow valves

4. Patented deformable tangs secure element in bowl

5. Coreless element assembly

6. Re-usable element support core

**Fig. 6.48- Sectional view of Pressure Line Filter WPF Series
(Courtesy of Parker)**

64

64

Technical Specifications

Mounting Method	4 mounting holes
Port Connection	
30	SAE-8, 1/2" NPT, 1/2" BSPP
60/110	SAE-12, 3/4" NPT, 3/4" BSPP
	3/4" SAE, Code 62
160/240/280	SAE-20, 1 1/4" NPT, 1 1/4" BSPP
	1 1/4" SAE, Code 62
330/660/1320	SAE-24, 1 1/2" NPT, 1 1/2" BSPP
	2" SAE Flange Code 62
Flow Direction	Inlet: Side Outlet: Side
Flow Capacity	
30	8 gpm (30 lpm)
60	16 gpm (60 lpm)
110	29 gpm (110 lpm)
160	42 gpm (160 lpm)
240	63 gpm (240 lpm)
280	74 gpm (280 lpm)
330	87 gpm (330 lpm)
660	174 gpm (660 lpm)
1320	200 gpm (757 lpm)
Housing Pressure Rating	
Max. Allowable Working Pressure	6090 psi (420 bar)
Fatigue Pressure	6090 psi (420 bar) @ 1 million cycles
Burst Pressure 30	15950 psi (1100 bar)
60/110	17400 psi (1200 bar)
160/240/280	17110 psi (1180 bar)
330/660/1320	15080 psi (1040 bar)
Element Collapse Pressure Rating	
BH4HC, V	3045 psid (210 bar)
ON, W/HC	290 psid (20 bar)
Fluid Temp. Range	14°F to 212°F (-10°C to 100°C)
Consult HYDAC for applications operating below 14°F (-10°C)	
Fluid Compatibility	
Compatible with all hydrocarbon based, synthetic, water glycol, oil/water emulsion, and high water based fluids when the appropriate seals are selected.	
Indicator Trip Pressure	
ΔP = 29 psid (2 bar) -10% (optional)	
ΔP = 72 psid (5 bar) -10% (standard)	
ΔP = 116 psid (8 bar) -10% (optional/ non bypass)	
Bypass Valve Cracking Pressure	
ΔP = 43 psid (3 bar) +10% (optional)	
ΔP = 87 psid (6 bar) +10% (standard)	
Non Bypass Available	

DF Series

Inline Filters

6090 psi • up to 200 gpm

**Fig. 6.49- Example of Line-Mounted
Top-Ported Pressure Filter
(Courtesy of Hydac)**

65

65

❖ **Pressure Drop:**
- Overall $\Delta p = \Delta p$ (housing) + Δp (element)
- Filter micron size $\uparrow \rightarrow \Delta p$ (element) $\downarrow$
- Filter mesh size $\uparrow \rightarrow$ micron size $\downarrow \rightarrow \Delta p$ (element) $\uparrow$

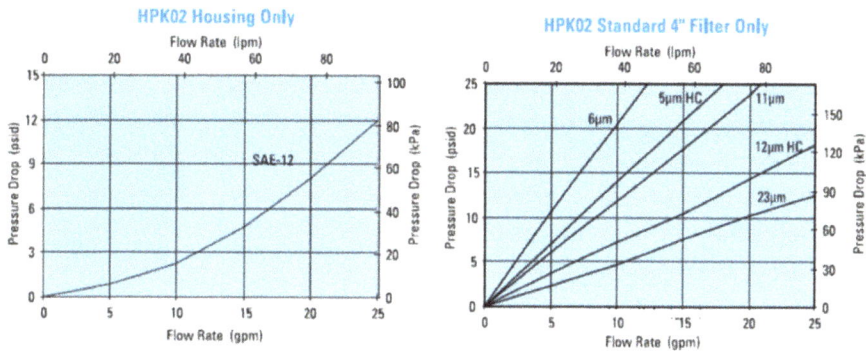

Fig. 6.50- Example of Line-Mounted Top-Ported Pressure Filter
(Courtesy of Hydac)

66

6.9.4- Last Chance Filters

❖ **Primary Duty:**
- Abrasive particles pass through main system filters →
- Last Chance Filters are used to protect critical components.
- Last Chance Filters are not intended to replace the system filter.

❖ **Placement:**
- Because they are placed in series.
- They are exposed to system pressure →
- They have a high collapse pressure.

❖ **Applications:**
- Servo circuits, precision machinery,
- Variable-displacement pump and motor systems,
- Hydrostatic drives, and machinery in dirty or dusty environments.

67

CP-C16 Series

Circuit Protector Manifold Cartridge Filters
3000 psi • up to 12 gpm

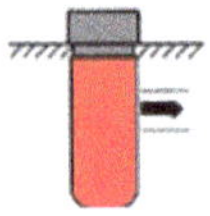

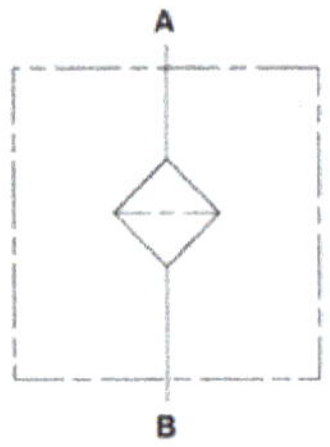

Technical Specifications

Mounting Method	C16-2 Cavity (SAE-16 Threaded Port)
Flow Direction	Inlet: Bottom Outlet: Side
Construction Materials	Steel
Flow Capacity	12 gpm (45 lpm)
Housing Pressure Rating	
Max. Allowable Working Pressure	3000 psi (210 bar)
Fatigue Pressure	Contact HYDAC Office
Burst Pressure	Contact HYDAC Office
Element Collapse Pressure Rating	
W/HC	250 psid (17 bar)
Fluid Temperature Range	14°F to 212°F (-10°C to 100°C) Consult HYDAC for applications operating below 14°F (-10°C)
Fluid Compatibility	
Compatible with all petroleum oils rated for use with Nitrile rubber (NBR) seals.	

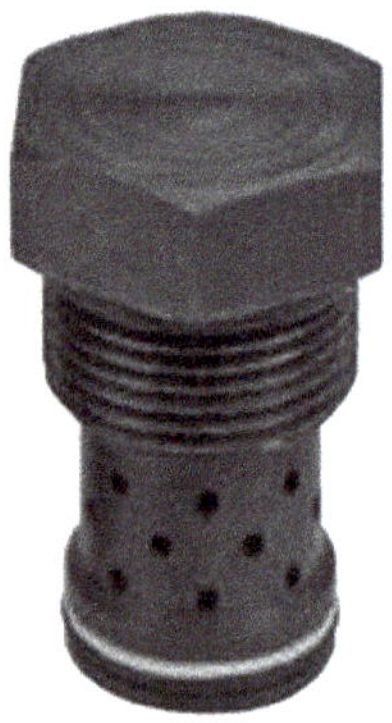

Fig. 6.51- Example of Last Chance Filters (Courtesy of Hydac)

68

68

6.9.5- Return Filters

❖ **Placement:**

- Open Circuits: Return filters are placed on the main return line.
- Closed Circuits: Return filters are placed on the case drain line.

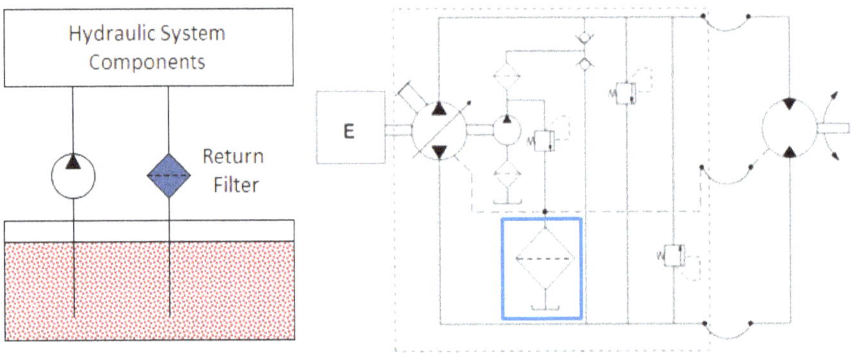

Open Circuit　　Closed Circuit

Fig. 6.52 – Placement of Return Filters

69

69

Technical Specifications

Mounting Method

75/90/150/165/185	2 mounting holes - filter housing
50/75/90/150/165/185/210/270/ 330/500/661/851/975/1100	4 mounting holes - filter housing

Flow Capacity

50 - 13 gpm (50 lpm)	270 - 71 gpm (270 lpm)
75 - 20 gpm (75 lpm)	330 - 87 gpm (330 lpm)
90 - 24 gpm (90 lpm)	500 - 132 gpm (500 lpm)
150 - 40 gpm (150 lpm)	661 - 174 gpm (660 lpm)
165 - 43 gpm (165 lpm)	851 - 225 gpm (850 lpm)
185 - 49 gpm (185 lpm)	975 - 258 gpm (950 lpm)
210 - 55 gpm (210 lpm)	1100 - 300 gpm (1100 lpm)

Housing Pressure Rating

Max. Allowable Working Pressure*	145 psi (10 bar), 101.5 psi (7 bar) *(Sizes 975 & 1100)*
Fatigue Pressure	145 psi (10 bar) @ 1 million cycles
Burst Pressure	75-500 >580 psi (40 bar)
	50, 661/851 536 psi (37 bar)
	975/1100 Consult Factory

Element Collapse Pressure Rating

BN4HC *(size 50, 975 & 1100 only)*	145 psid (10 bar)
ON *(size 50-851 only)*, W/HC	290 psid (20 bar)
ECON2, BN4AM, AM, P/HC, MM	145 psid (10 bar)
V	435 psid (30 bar)

Fluid Temperature Range -22°F to 212°F (-30°C to 100°C)

Consult HYDAC for applications below -22°F (-30°C)

Fluid Compatibility

Compatible with all hydrocarbon based, synthetic, water glycol, oil/water emulsion, and high water based fluids when the appropriate seals are selected.

Indicator Trip Pressure

P = 20 psi (1.4 bar) - 10%
P = 29 psi (2 bar) -10% *(standard)*
P = 72 psi (5 bar) -10% *(optional)*

Bypass Valve Cracking Pressure

ΔP = 43 psid (3 bar) +10% *(Standard - All sizes except 50, 975, 1100)*
ΔP = 87 psid (6 bar) +10% *(Optional - Sizes 50, 975 & 1100 not available)*
ΔP = 25 psid (1.7 bar) +10% *(Standard for Sizes 50, 975 & 1100)*

RFM Series
In-Tank Return Line Filters
145 psi • up to 224 gpm

Fig. 6.53- Example of Return Filters (Courtesy of Hydac)

70

6.9.6- Combined Return and Suction Booster Filter

❖ **Description:**

- Many hydraulic circuits may have more than one operating pump.
- A return filter is used for the open circuit.
- A suction filter is used for the boosting pump in the closed circuit.

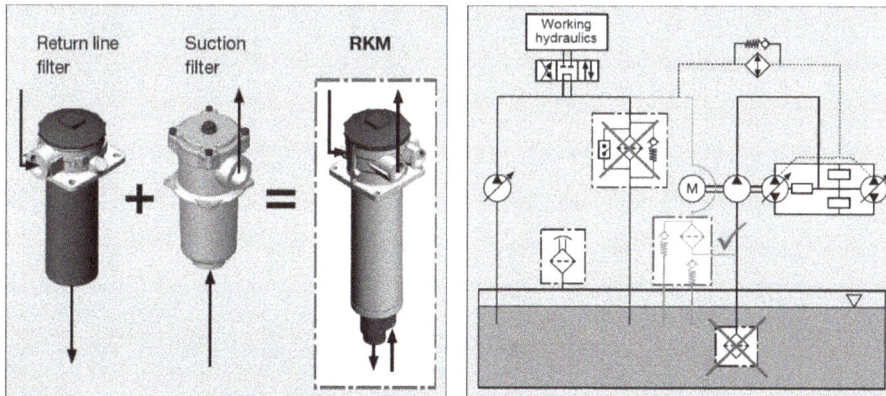

Fig. 6.54- Combined Return and Suction Booster Filter (Courtesy of Hydac)

71

❖ **Advantage:**

- Cost and space saving.

- Easy maintenance.

- Meets automotive standard.

- Standard Connections: pipe, SAE straight thread, flange.

- Standard Ports: ISO 228 porting, NPTF inlet and outlet female test ports.

- Available with magnet inserts.

- Various Dirt Alarm options.

- Available with housing drain plug.

❖ **Typical Applications:**

- Machines with two or more circuits,

  o Mobile machines (wheel loaders, forklifts).

  o automotive engineering.

72

72

More Info about Combined Return and Suction Filter review in textbook:

- **Figure 6.55**

73

73

6.9.7- Diffusers

- Concentric tubes designed with discharge holes →
- fluid (aeration, foaming and noise) ↓ →
- Better system performance and longer component life.

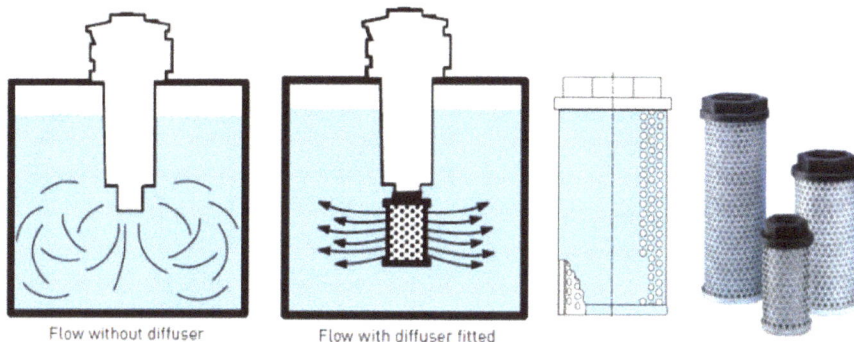

Flow without diffuser   Flow with diffuser fitted

Fig. 6.56- Diffusers on a return line (Courtesy of Parker)

74

74

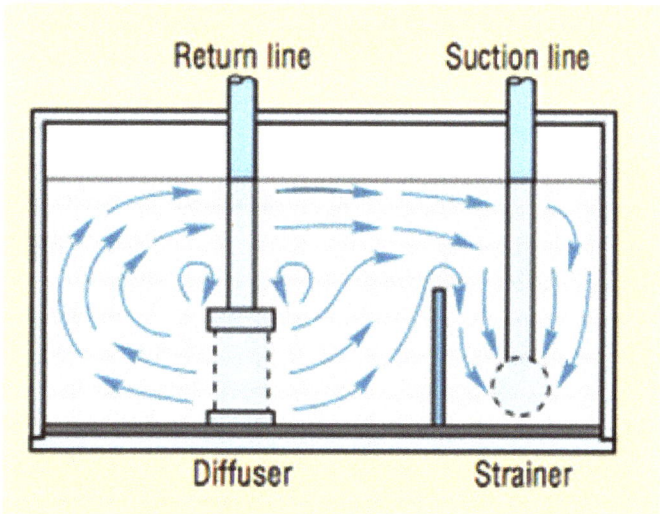

**Fig. 6.57- Flow Streams from the Return Line to Suction Line
(Courtesy of Parker)**

75

75

6.9.8- Filler Caps

- **Sealed Cover** → prevent ingression of contaminants when closed.
- **Filling Screen** → catch relatively large contaminants during filling.
- **Filler Cap** → chained to the reservoir to keep it captive.
- **Lock (optional)** → more protection.

Video 723

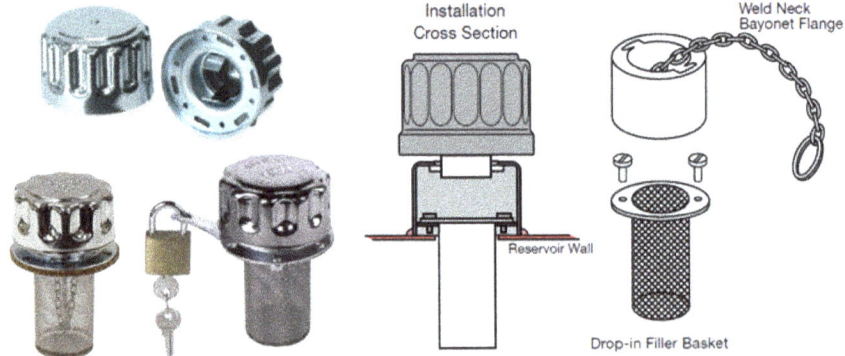

Installation Cross Section

Weld Neck Bayonet Flange

Reservoir Wall

Drop-in Filler Basket

Fig. 6.58- Filler Caps

76

76

- **Breathing** → allows reservoir breathing (without filter breather)
1. Air intake to reservoir through vacuum breaker (0.435 psi)
2. Air venting to atmosphere through relief valve (5 or 10 psi).

More Info about Filling Caps

- **Figure 6.59 through 6.65**

77

77

6.9.10- Offline (Bypass) Filtration Units

❖ **Placement:** portable unit or permanently installed unit.

❖ **Primary Duty:**
▪ Filters the full volume of oil in the reservoir.
▪ It does not subject to surge flows.
▪ Filter element is replaced without interrupting the system.

❖ **Cost:**
▪ High initial cost → justified by extended system life.

❖ **Sizing:**
▪ Size is function of (reservoir volume + pump flow + # circulations/Hr).
▪ Example:
  o Reservoir volume = 100 liters.
  o Filtration Pump Q = 5-100 liter/min →
  o Reservoir is filtered in 10-20 minutes →
  o Reservoir is filtered 3-6 times per hour.

78

78

❖ **Filtration Rating:**
▪ Up to < 2 microns are possible.
▪ Water-absorbent filters and heat exchangers can be included.

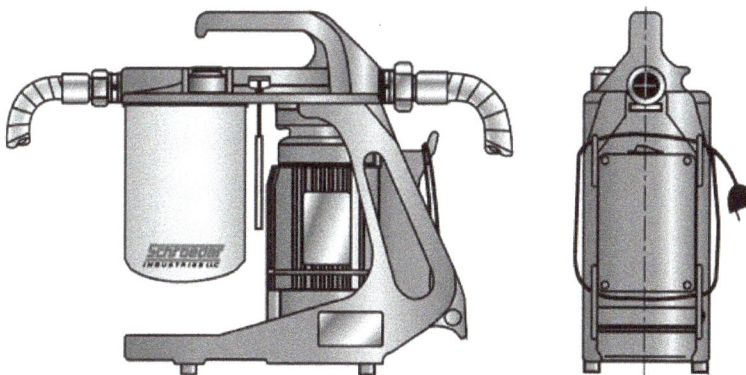

**Fig. 6.66 – Example of Hand-Portable Offline Filtration Unit
(Courtesy of Schroeder)**

79

79

kidney loop filtration

Video 480 (3 min)

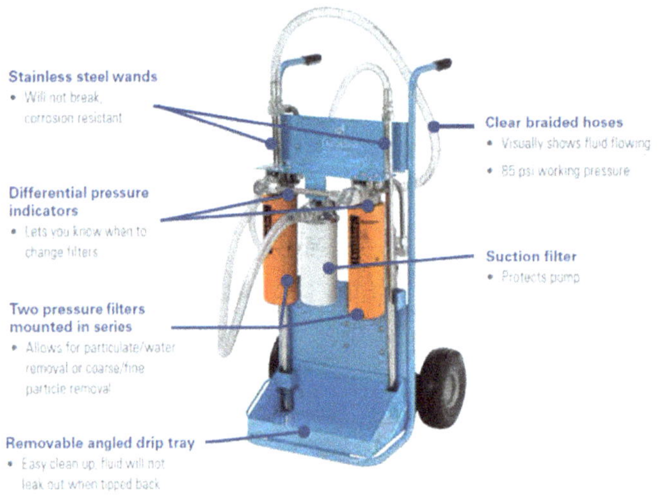
Stainless steel wands
• Will not break, corrosion resistant
Differential pressure indicators
• Lets you know when to change filters
Two pressure filters mounted in series
• Allows for particulate/water removal or coarse/fine particle removal
Removable angled drip tray
• Easy clean up, fluid will not leak out when tipped back
Clear braided hoses
• Visually shows fluid flowing
• 85 psi working pressure
Suction filter
• Protects pump

Fig. 6.67 – Example of Cart-Portable Offline Filtration Unit (Courtesy of Donaldson)

80

80

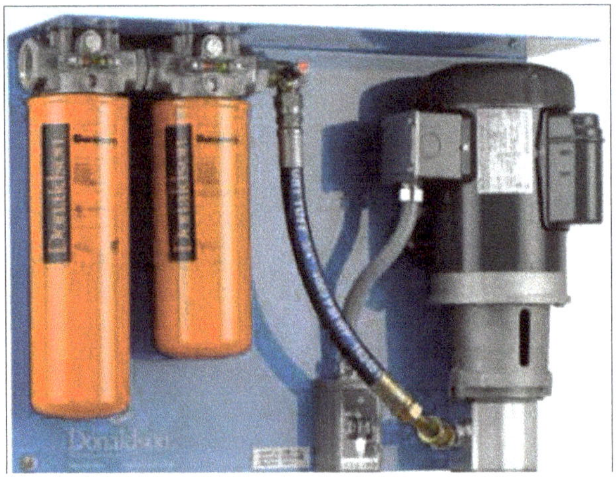
Donaldson
Donaldson
Donaldson

Fig. 6.68 – Example of Fixed-Mounted Offline Filtration Unit (Courtesy of Donaldson)

81

81

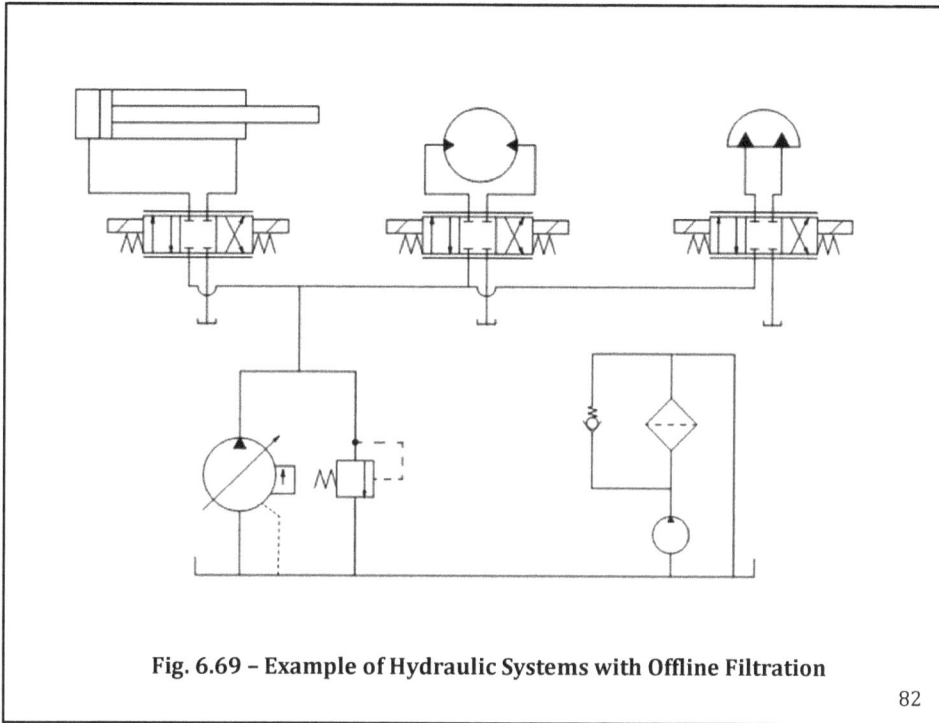

Fig. 6.69 – Example of Hydraulic Systems with Offline Filtration

82

82

Chapter 6 Reviews

1. Hydraulic filters are classified based on?
    A. Application.
    B. Place in a hydraulic circuit.
    C. Body style and Types of contamination.
    D. All the above.

2. In a hydraulic filter oil flows:
    A. From outside to inside of filter media
    B. From inside to outside of filter media
    C. Parallel to the filter media inside the filter housing
    D. Maximum heating capacity of 8 W/cm^2.

3. A duplex hydraulic filter is used to:
    A. Increase the dirt holding capacity of the filtration system.
    B. Improve the cleanliness and filtration efficiency of the system.
    C. Reduced the pressure drop across the filter media.
    D. Replace the filter without interrupting the machine operation.

4. A dual hydraulic filter is used to:
    A. Increase the dirt holding capacity of the filtration system.
    B. Improve the cleanliness and filtration efficiency of the system.
    C. Reduced the pressure drop across the filter media.
    D. Replace the filter without interrupting the machine operation.

5. A staged filtration system is used to:
    A. Increase the dirt holding capacity of the filtration system.
    B. Improve the cleanliness and filtration efficiency of the system.
    C. Reduced the pressure drop across the filter media.
    D. Replace the filter without interrupting the machine operation.

Chapter 6 Assignment

Student Name: -- Student ID: ------------------

Date: -- Score: -----------------------

Assignment:

Draw a hydraulic circuit diagram showing possible locations for placing hydraulic filters.

Chapter 7
Filter Media and Filtration Mechanisms

Objectives:

This chapter presents an overview of filter elements including the construction and material of the filter media. This chapter discusses surface filters versus depth filters. The chapter discusses also the principles of various filtration mechanisms that are applicable in hydraulic filters such as direct interception, absorption, adsorption, and magnetic separation.

Brief Contents:

7.1- Filtration Mechanisms

7.2- Materials for Filter Media

7.3- Filter Media Structure

0

0

7.1- Filtration Mechanisms

❖ **Retaining Large Size Particles Mechanically by Inertia:**

▪ Fluid is accelerated OR changes direction between fibers →

▪ Large and heavy particles → large inertia → slower motion →

▪ Particles trapped and retained by the media fibers.

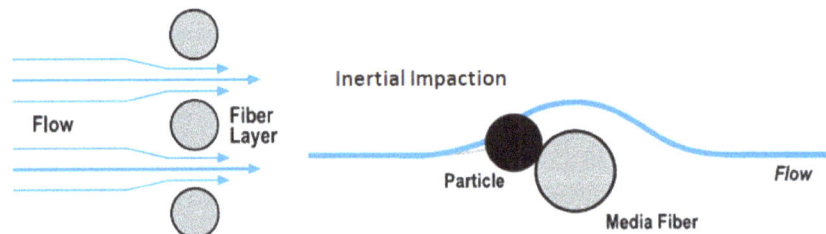

Fig. 7.1- Retaining Large Size Particles by Inertia (Courtesy of Donaldson)

1

1

❖ **Retaining Medium Size Particles Mechanically by Direct Interception:**

▪ Direct interception s also referred as "Sieving".

▪ Mid-size particles (neither large inertia nor small to diffuse) →

▪ But > filter micron size → mechanically captured and retained by media

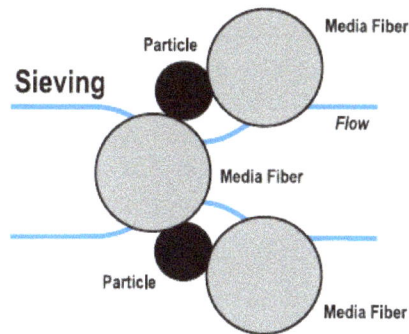

**Fig. 7.2- Retaining Medium Size Particles by Direct Interception
(Courtesy of Donaldson)**

2

2

❖ **Retaining Small Size Particles by Absorption:**

▪ Also referred to as Diffusion.

▪ Fiber media → electrostatic forces → molecular attraction →

▪ Small particles are attracted and collected by the fiber.

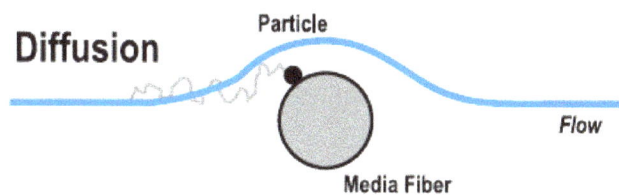

**Fig. 7.3- Retaining Small Size Particles by Diffusion
(Courtesy of Donaldson)**

3

3

Other Mechanisms:

❖ **Retaining Particles by Adsorption:**
- Chemically treated filter media → chemical removal of contaminants.
- Caution: desirable additives my be removed.

❖ **Oil Cleaning by Centrifugal Separators:**
- Water and solid particles with a density > oil density →
- Centrifugal Separators (explained in Volume 3) →
- Separation of water and contaminants.
- Caution: water can't be removed 100% + possible additive removal.

❖ **Oil Cleaning by Vacuum Dehydration:**
- Vacuum Filters (explained in Volume 3) →
- Water removal.

❖ **Oil Cleaning by Magnetic Separation:**
- Magnetic Separation useful in separation of ferrous solids from fluid streams.

4

4

7.2- Materials for Filter Media

❖ **Cellulose Fibers Filter Media (Traditional):**
- **Material:** Wooden fibers held together by resin.
- **Shape:** Irregular shape.
- **Size:** Irregular small (microscopic) pores size.
- **Flow and ΔP:** High flow resistance → high Δp.
- **Filtration:**
  - Good through the depth of the media.
  - Poor filtration performance as compared to synthetic media.
- **Fluid:** wide variety of petroleum-based fluids.

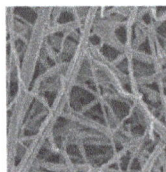 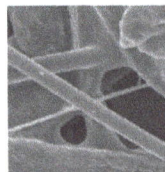  

SEM 100X   SEM 600X   MEDIA IMAGE   HOW IT WORKS

Fig. 7.4- Cellulose Fibers Filter Media (Courtesy of Donaldson)

5

5

❖ **Synthetic Fibers Filter Media (Fully Synthetic):**

▪ **Material:** Man-made, smooth, rounded fibers.

▪ **Shape:** Consistent shape.

▪ **Size:** Controlled size and distribution pattern through the media.

▪ **Flow and ΔP:** Low flow resistance → low Δp.

▪ **Filtration:**

▪ Consistency of fiber shape → dirt catching on both (surface & depth) → dirt holding capacity ↑.

▪ **Fluid:** Ideal for synthetic fluids, water glycols, water/oil emulsions, HWCF and petroleum-based fluids.

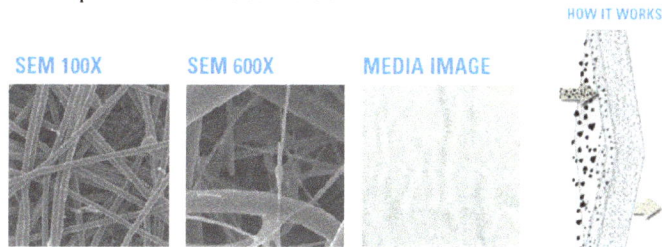

Fig. 7.5- Synthetic Fibers Filter Media (Courtesy of Donaldson)

6

6

❖ **Combined Fibers Filter Media (Cellulose & Synthetic):**

▪ Effective fuel filtration performance for optimal protection.

Fig. 7.6- Combined Fibers Filter Media (Courtesy of Donaldson)

7

7

❖ **High Performance Synthetic Fibers Filter Media:**
- Today's fluid requirements: (fire resistance, biodegradability, chemical and thermal resistances, and electrical insulating ability) →
  High Performance Synthetic Fibers filter media.
- **Material:** A blend Glass Fibers bonded with epoxy-based resin system.
- **Flow and ΔP:** High flow resistance → high Δp.
- **Fluid:** Ideal for phosphate ester and water glycol fluids providing.
- **Filtration:** Ideal for fine filtration and precision components.

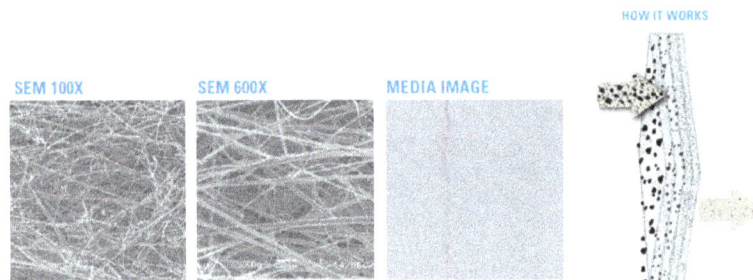

**Fig. 7.7- High Performance Synthetic Fibers Filter Media
(Courtesy of Donaldson)**

8

8

❖ **Wire Mesh Filter Media:**
- **Material:** Stainless steel, epoxy-coated wire mesh
- **Shape:** Consistent shape.
- **Size:** Available in different sizes.
- **Flow and ΔP:** Least flow resistance → lowest pressure drop.
- **Filtration:**
  - Available in various micron sizes and ranging from 100 to 500 microns.
  - Catch large and harsh particles.
  - Generally used in suction strainers.

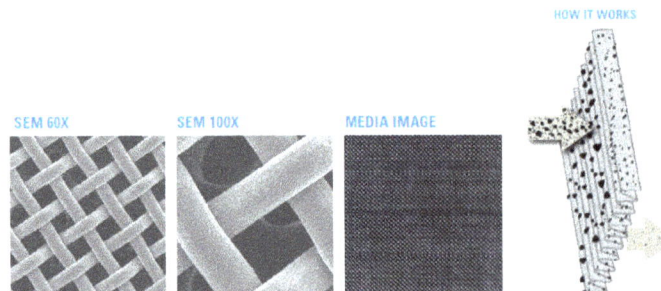

Fig. 7.8- Wire Mesh Filter Media (Courtesy of Donaldson)

9

9

❖ **Water Absorption Filter Media:**

▪ **Material:** super-absorbent polymer →

▪ Quickly and effectively removes free water from hydraulic systems →

▪ Used for petroleum-based fluids.

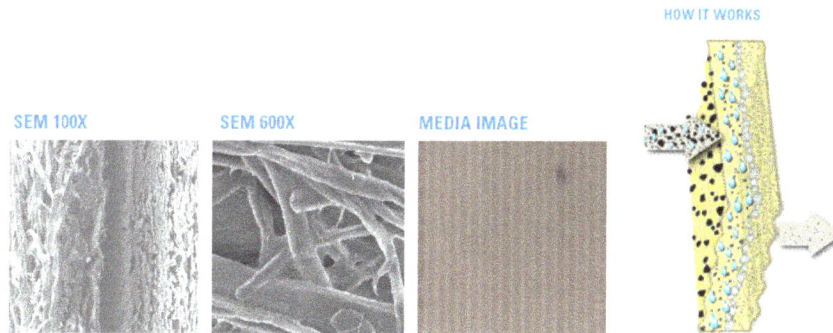

**Fig. 7.9- Water Absorption Filter Media
(Courtesy of Donaldson)**

10

10

7.3- Filter Media Structure

▪ **Fiber Structure (1):** → fluid accelerate + change direction → retention

▪ **Uniform vs. Graded Pore Size (2):** Graded → dirt holding capacity ↑+ ΔP ↓.

▪ **Fixed vs. Non-Fixed Pore Size (3):** Better bonding → fixed pore media → resist surges of (P, T, and Q) → no media migration.

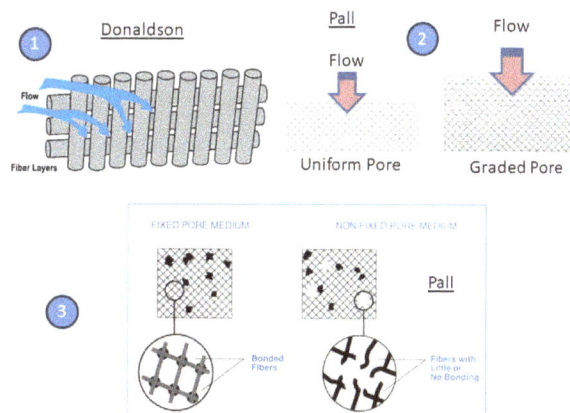

**Fig. 7.10- Filter
Media Fibers**

11

11

- **Surface vs. Depth:** filter media structures.

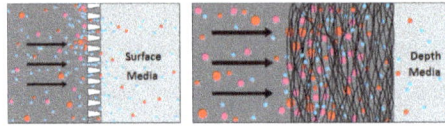

Fig. 7.11- Surface versus Depth Filter Media (Courtesy of Bosch Rexroth)

7.3.1- Surface Filter Media

 Video 243 (0.5 min)

❖ **Filtration Process:**
- Primary filtration mechanism is direct interception.
- Soft and deformable particles intrusion →
- Media become partially blocked → Pore size overtime ↓
- → filter is completely clogged.

❖ **Applications:** strainers, suction filters, and lubricating systems.

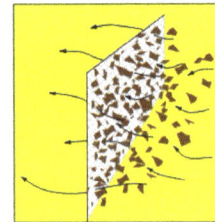

Fig. 7.12- Filtration using Surface Filters

12

12

❖ **Advantages of Surface Filter Media:**
- Are washable and cleanable.
- No media migration with the oil.
- Low flow resistance and Δp.
- High fatigue and corrosion resistant.
- Work at high temperature.

❖ **Disadvantages of Surface Filter Media:**
- Catch only relatively large contaminants.
- Can't be used to maintain high cleanliness Level.
- Needle-shaped contaminants ($D <$ Pore size and $L >$ pore size) can still pass through these filters.

❖ **Material:**
- Stainless steel, galvanized iron, or phosphor bronze.
- Ribbon-shaped metal.
- Accordion-pleated paper.
- Stacked metal disks.

 Video 666 (1 min)

13

13

❖ **Structure of Metallic Surface Filters:**

▪ Square are most common.

▪ Braided → better filter rating.

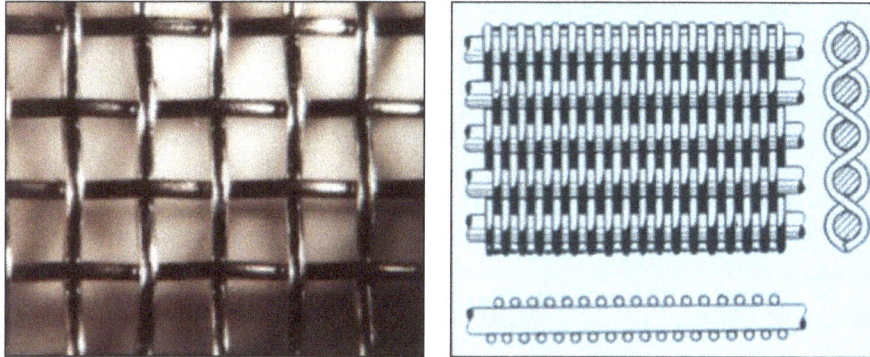

**Fig. 7.13- Square versus Braided Wire Mesh Surface Filters Media
(Courtesy of Bosch Rexroth)**

14

14

7.3.2- Depth Filter Media

❖ **Filtration Process:**
▪ Classified as Absorbent filters.
▪ Direct Interception (med size) and Absorption (small size).

❖ **Applications**: pressure, return, and offline filters.

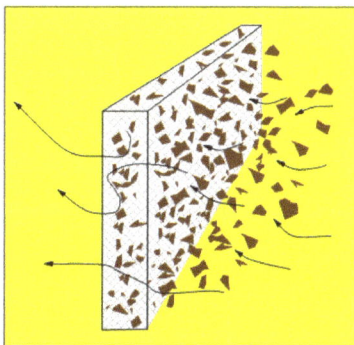

📹 Video 244 (0.5 min)

Fig. 7.14- Filtration using Depth Filters

15

15

❖ **Advantages of Surface Filter Media:**
- Effective filtration for small contaminants.
- Used to maintain high cleanliness Level.
- Has large dirt holding capacity.

❖ **Disadvantages of Surface Filter Media:**
- Are not washable or cleanable.
- Possible media migration with the oil.
- High flow resistance and pressure drop.

❖ **Material:**
- Various materials:
- Organic (such as Cellulose, or Cotton).
- Synthetic fibers.
- Randomly oriented steel wires.
- Must be compatible with working (fluid & Temperature).

16

16

❖ **Structure:**
- Layers of fibers (D = 0.5 to 30 microns).
- Fibers are wounded in layers (layer depth = 0.25 – 2 mm "0.01-0.08 in").
- No 100% consistent pore size → rated based on average pore size.
- Quality of depth filter elements depends on:
  - ○ Consistency of pore size through the media.
  - ○ Fiber bonding → ability to prevent media migration.
  - ○ Central support of the filter element pleats.
  - ○ Sealing to end caps.

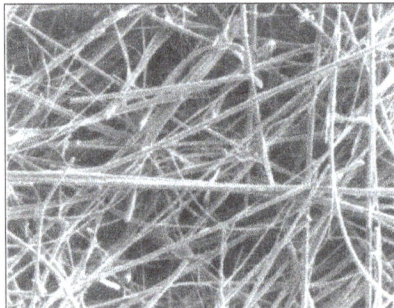

Video 667 (1 min)

Fig. 7.15- Structure of Depth Filters Media

17

17

❖ **Example 1– Conventional Glass Fiber Pressure Filter:**

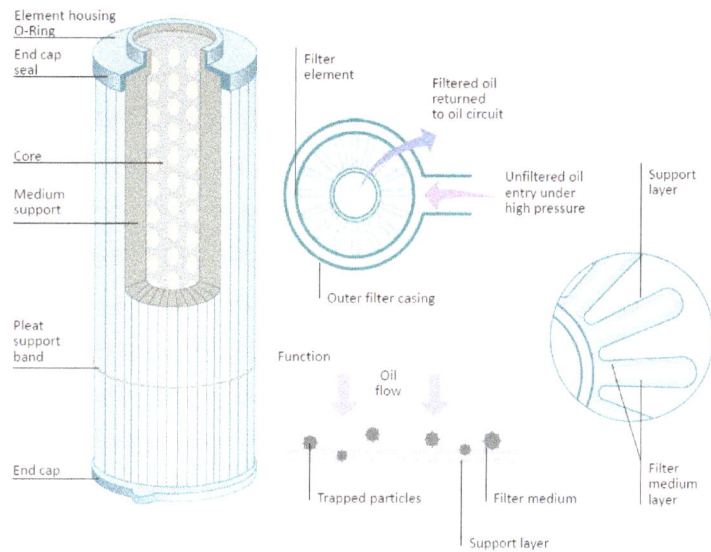

Fig. 7.16- Example of Glass Fiber Pressure Filter (Courtesy of C.C. Jensen Inc.) 18

18

❖ **Features**

- Glass Fiber → good for pressure filter (high **p** + high **Q** + little **Δp**).

- Glass Fiber → filtration rating of 5 – 50 microns.

- Glass Fiber → remove solid particles only.

- Glass Fiber → small depth → limited dire holding capacity (1–100 grams).

- Pressure shocks at stop/start → captured particles will be released again.

19

❖ **Example 2– Conventional Micro Glass Fiber Pressure Filter:**

Each layer performs a necessary function

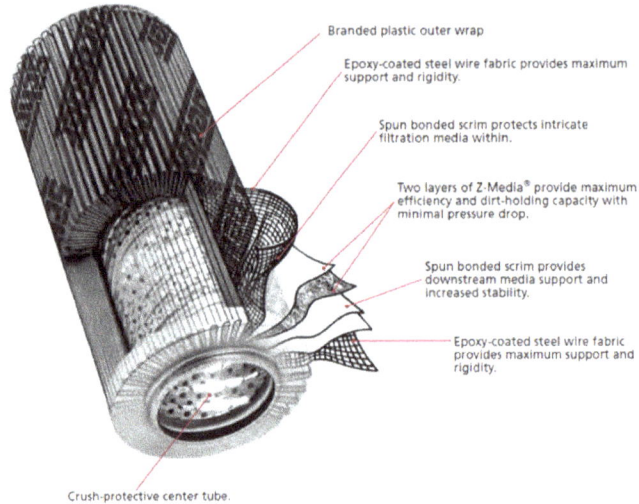

Branded plastic outer wrap

Epoxy-coated steel wire fabric provides maximum support and rigidity.

Spun bonded scrim protects intricate filtration media within.

Two layers of Z-Media® provide maximum efficiency and dirt-holding capacity with minimal pressure drop.

Spun bonded scrim provides downstream media support and increased stability.

Epoxy-coated steel wire fabric provides maximum support and rigidity.

Crush-protective center tube.

Fig. 7.17- Example of Glass Fiber Pressure Filter "Z-Media®"
(Courtesy of Schroeder)

20

20

❖ **Examples 3– Laid-Over Pleating in Depth Filter Element Technology:**

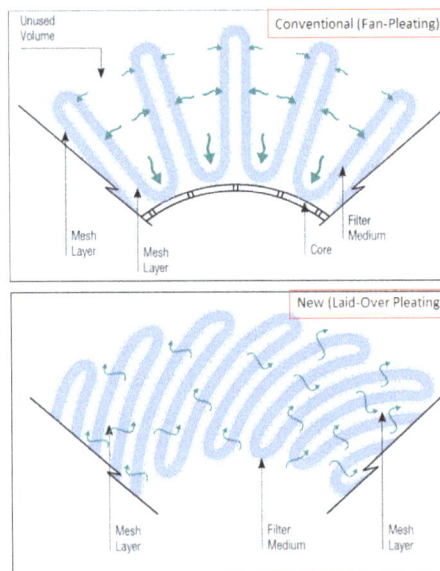

Video 468 (1 min)

Video 469 (2 min)

Video 470 (1 min)

- More filtration area for same size ↑.
- Uniform flow distribution.
- Protects pleat against collapse and bunching.
- Anti-static → minimizes static discharges.
- Resistance to cyclic flow and pressure.

Fig. 7.18- Example of New Depth Filter Element Technology "Ultipleat"
(Courtesy of Pall)

21

21

❖ **Example 4– New Depth Filter Element Technology:**

▪ An outer Rap (1) → Media robustness ↑.

▪ New pleating shape (2) → Flow surface ↑ + flow velocity ↓ + Δp across the media ↓.

Fig. 7.20- Example of New Depth Filter Element Technology "Optimicron" (Courtesy of Hydac)

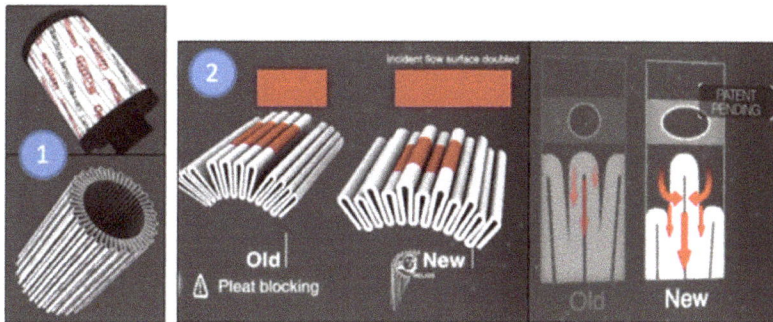

22

22

▪ 7 consecutive layers (3) → filtration effectiveness ↑

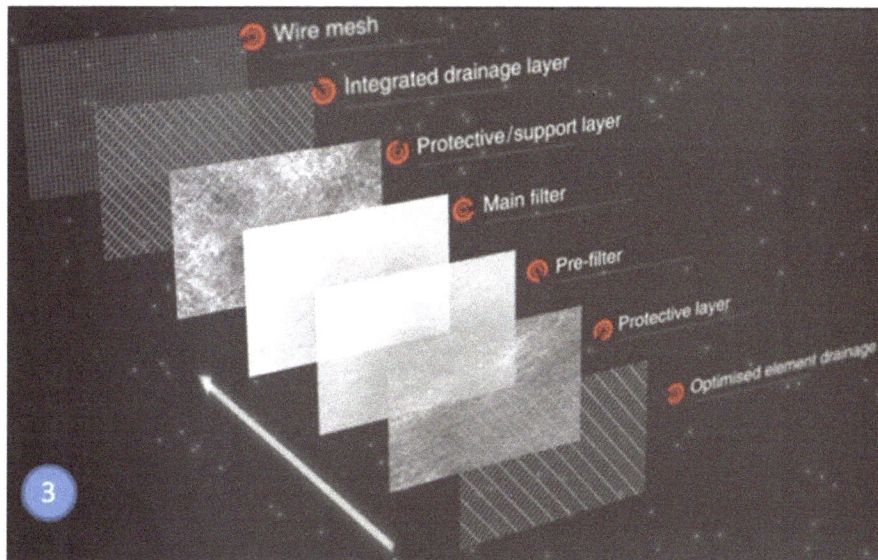

Fig. 7.20- Continue Video 417 (7 min) 23

23

❖ **Example 5– Synthetic Depth Filter:**
- High-efficiency filtration rating.
- Exceptionally low flow resistance
- Consistent performance throughout filter life.
- Excellent fluid compatibility.
- Ideally suited for: heavy-duty mobile equipment, in-plant hydraulics, transmissions, and bearing lube oil systems.

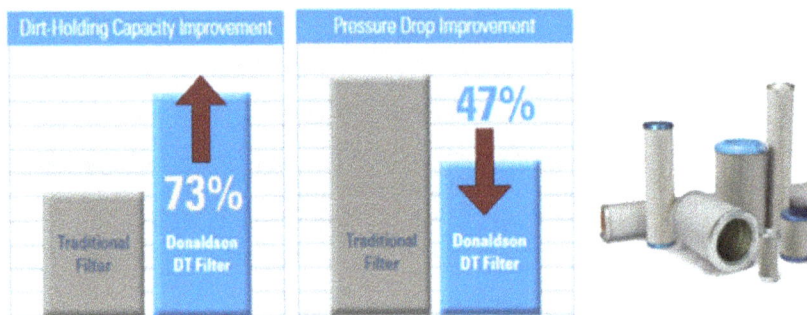

Fig. 7.21- Example of Synthetic Filter Element "DT Filters"
(Courtesy of Donaldson)

24

24

- **Epoxy-Coated Steel Support Mesh (1):**
  - Two layers at the upstream and downstream sides →
  - Excellent pleat support →
  - Protect the media against damage during handling and installation.

- **Media Support Layers (2):**
  - Two layers at the upstream and downstream sides →
  - Protect the media during pressure surges

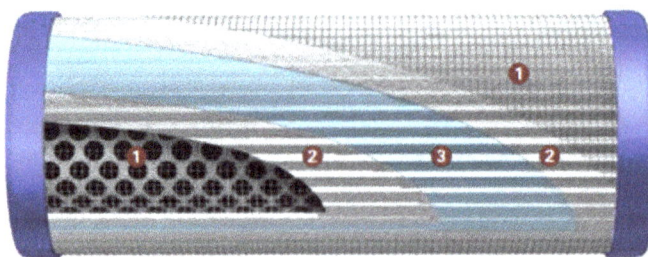

Fig. 7.21- Continue

25

25

- **Synteq™ Media Technology (3):**
  o Synthetic filter media has smooth, rounded fibers → Low Δp.
  o Ideal for filtering synthetic fluids, water glycols, water oil emulsions, HWCF (high water content fluids), and petroleum-based fluids.

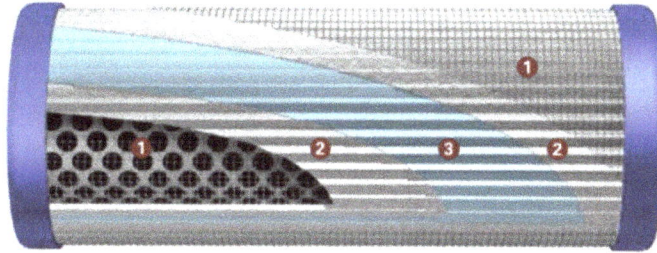

Fig. 7.21- Continue

26

26

❖ **Examples 6– Offline Filter for High Dirt Holding Capacity:**

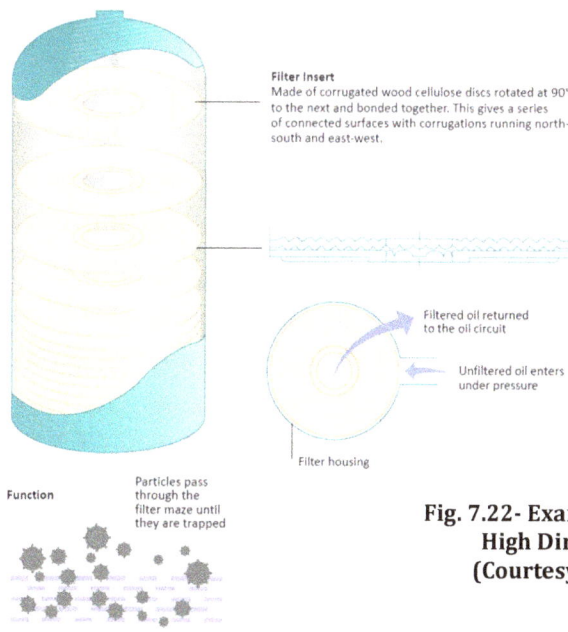

Filter Insert
Made of corrugated wood cellulose discs rotated at 90° to the next and bonded together. This gives a series of connected surfaces with corrugations running north-south and east-west.

Filtered oil returned to the oil circuit

Unfiltered oil enters under pressure

Filter housing

Function

Particles pass through the filter maze until they are trapped

Fig. 7.22- Example of Offline Filter for High Dirt Holding Capacity (Courtesy of C.C. Jensen Inc.)

27

27

❖ **Features:**

▪ Offline filters → large dirt holding capacity →

  ○ ≈ 4 liters solid, 2 liters water, and

  ○ ≈ 4 liters of oil degradation products (Varnish, Sludge, and Oxidation.

▪ Replaced ≈ only on annual bases.

▪ Such filters are good to filter particles as small as 3 microns.

28

❖ **Example 7- Water Removal Filer Element:**

▪ Specially designed Aquamicron® filter elements →

▪ Separate free water from mineral oils →

Technical Specifications

Collapse Rating	145 psid (10 bar)
Temperature range	32°F to 212°F (0°C to 100°C)
Compatibility with hydraulic media	Mineral oils: Test criteria to ISO 2943 Lubricating oils: Test criteria to ISO 2943 Other media available on request
Opening pressure of by-pass valves	ΔP0 = 43 psid ±7 psi (3 bar ±0.5 bar)
Bypass valve curves	The bypass valve curves apply to mineral oils with a specific gravity of 0.86. The differential pressure of the valve changes proportionally with the specific gravity.

Fig. 7.23- Example of Water Removal Filter Element "Aquamicron®"
(Courtesy of Hydac)

29

- A clogging indicator → monitor Δp

- A bypass valve → limit Δp across the filter element.

- Δp-Q for mineral oils with a SG = 0.86.

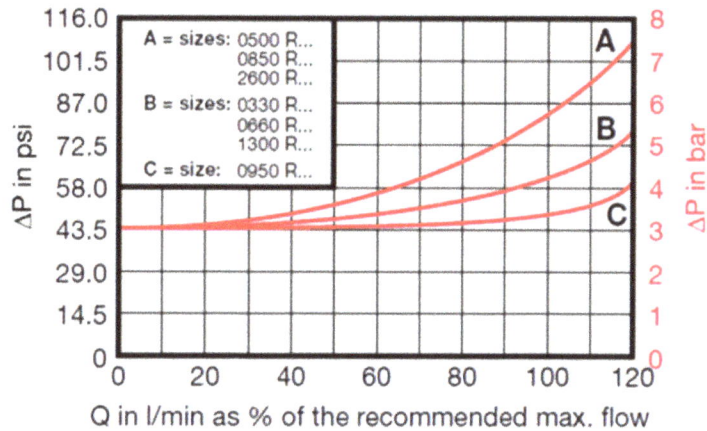

Fig. 7.23- Continue

30

30

❖ **Example 8- Water Removal Filer Elements:**

- Filter media swollen when it becomes wet with absorbed water.

- Effective in removing free water mineral-base and synthetic fluids.

- Table to calculate the water content in a specific volume of oil.

If you Have:	Multiply By:	To Get:
mg/l	0.00009	%
ppm	0.0001	%
ml	1.0	cc
cc	0.0338	fluid ounces
cc	0.00106	quarts
cc	0.000264	gallons

Fig. 7.24- Example of Water Removal Filter Element "Par-Gel"
(Courtesy of Parker)

31

31

- ▪ Example:

  ○ Reservoir volume = 200 gallon and Water contaminated (1000 ppm) →

  ○ %Water content 1000 x 0.0001 = 0.1% of the oil volume →

  ○ Water Volume = (0.1 x 200)/100 gallon = 0.2 gallons of water.

  ○ If the acceptable water content in the oil = 300 ppm →

  ○ %Water that should be removed = 700 ppm = 700 x 0.0001 = 0.07 % →

  ○ Water volume that should be removed = (0.07x200)/100 = 0.14 gallons.

  Correction in Textbook (if needed)

32

32

- ❖ **Example 9– Oil Cleaning by Magnetic Separation:**

  ○ Magnetic Drain Plug (1):
  ✓ The most basic type of magnetic filter
  ✓ It should be periodically removed, inspected and wiped.
  ✓ Commonly used in engine oil pans, gearboxes and occasionally in hydraulic reservoirs.

  ○ Magnetic Rods (2):
  ✓ Magnetic Rods are immersed inside the reservoir.
  ✓ Magnetic rods can hold more particles than the drain plugs.

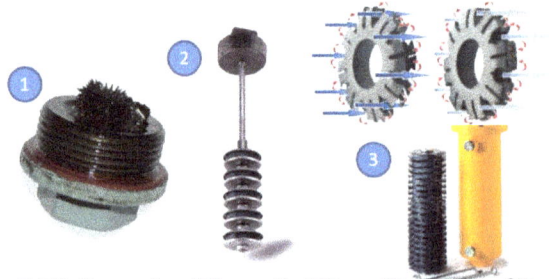

Fig. 7.25- Example of Magnetic Filters (Courtesy of Noria)

33

33

o Flow-through Magnetic Filters (3):
✓ As fluid passes through the slots →
✓ ferromagnetic particles accumulate in the gap between the plates.
✓ Cleaning process:
• Remove the filter core
• Blow the debris out from between the collection plates with an air hose.
• Wipe and clean.

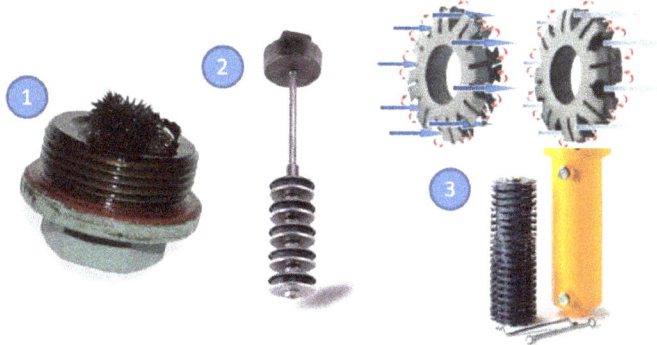

Fig. 7.25- Continue

34

34

❖ **Example 10– Combo Mechanical and Magnetic Filter:**
▪ Spin-on mechanical filter with steel housing (bowl).
▪ Magnetic wraps at the exterior wall of the housing.
▪ High power magnet at the bottom.
▪ Ferromagnetic debris are held tightly against the internal surface and at the bottom.

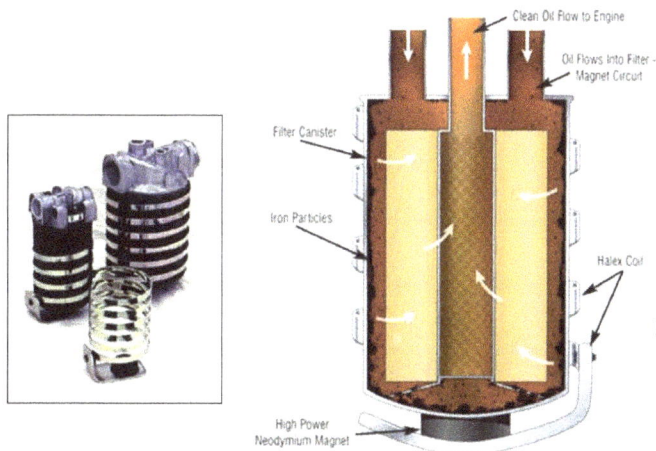

Fig. 7.26- Example of Mechanical and Magnetic Filters (Courtesy of Noria)

35

35

Chapter 7 Reviews

1. Which of the following filtration mechanisms is adequate for removing large sized particles?
   A. Absorption
   B. Adsorption
   C. Direct Interception
   D. Inertia

2. Which of the following filtration mechanisms is adequate for removing medium sized particles?
   A. Absorption
   B. Adsorption
   C. Direct Interception
   D. Inertia

3. Which of the following filtration mechanisms is adequate for removing small sized particles?
   A. Absorption
   B. Adsorption
   C. Direct Interception
   D. Inertia

4. Which of the following filtration mechanisms is adequate for removing Varnish?
   A. Absorption
   B. Adsorption
   C. Direct Interception
   D. Inertia

5. Which of the following filters is adequate for removing moisture?
   A. Pressure filters
   B. Return filters
   C. Desiccant filters
   D. Suction filters

Chapter 7 Assignment

Student Name: --- Student ID: -------------------

Date: -- Score: -----------------------

Assignment: Explain the benefits of laid over pleating technology over regular pleating in depth filters.

Chapter 8
Filter Selection Criteria

Objectives:

This chapter presents a selection checklist as a guide for selecting proper filters. The chapter also discusses briefly the concepts for cost-effective filtration and selecting a filter cleanliness level based on system requirements. This chapter presents several examples of filtration solution for hydraulic systems.

Brief Contents:

8.1- Filter Selection Checklist

8.2- Cost-Effective Filtration

8.3- Filter Selection Based on Cleanliness Requirements

8.4- Examples of Filtration Solutions   0

0

8.1- Filter Selection Checklist

❖ Questions in Filter Selection Checklist →

- Filter Purpose:
  - o For regular operation, for flushing, for water or varnish removal?
  - o For mobile or industrial application?

- Filter Location:
  - o Inline filter (suction -pressure – return), or offline filter?

 Video 668 (4 min)

- System Operating Conditions:
  - o What is the maximum system P, T, and Q?
  - o Are there possible pressure fluctuation and/or pressure spikes?
  - o Are there possible flow fluctuation or flow surges?
  - o Are there sensitive components that are intolerant to contamination?
  - o What type of the hydraulic fluid is used and its viscosity?

1

- System Cleanliness Requirements:
  o What is the required absolute/nominal beta ratio and filter efficiency?
  o What is the mesh size (for screens)?
  o Requirements for bypass?
  o Requirements for clogging indicators/alarms/control?
  o What is the anticipated dirt holding capacity?

- Filter Media:
  o Collapse pressure and flow fatigue resistance?
  o Moisture absorbance characteristics?
  o Anti-Static characteristics?

- Filter Housing:
  o Burst pressure and fatigue pressure?
  o Body style (inside tank, line mounted, on-tank top)
  o Port size?

Video 458 (1.5 min)

2

2

8.2- Cost-Effective Filtration

❖ **Cost vs. Service Life:**
- Proper filtration →
- Overall system reliability ↑ →
- Overall running cost ↓.

❖ **Service Life Vs. Filter Area:**
- Filter element area ↑→
- Dirt holding capacity ↑→
- Time before clogged) ↑→
- Overall running cost ↓.

❖ **Example:**
- Filter element area ↑ (x3) →
- Filter lifetime ↑ (x4-6)

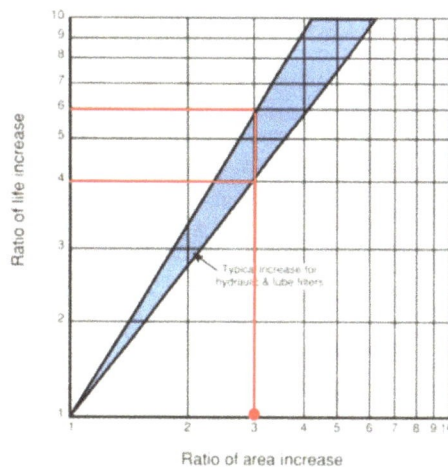

Fig. 8.1- Filter Element Service Life versus Filter Area

3

3

❖ **Staged Filtration:**
- Staged filtration (two or more filters in series)→
- Lower or finer filtration rating downstream →
- Compromise filter cost and system cost.

❖ **Filter Location and Cost:**
- Return filters or offline filters → Most cost-effective filtration solutions.
- Sensitive components (proportional, servo, EH pumps) →
- Pressure filters might be mandatory → most expensive.
- Compromise filter cost & components cost.

❖ **Bypass OR Non-Bypass:**
- Having a bypass → limits Δp
- → use filter element with less collapse pressure ↓ → cost ↓
- Sensitive components → Non-bypass or bypass-to-tank must be used.
- Compromise bypass valve with cost of filter element.

4

4

8.3- Filter Selection Based on Cleanliness Requirements

- User responsibility →

- Maintain cleanliness level as required by system manufacturer →

- Otherwise, system warrantee may be voided.

- If no recommendations were found →

- Follow general guidelines.

5

5

Application	Oil cleanliness required in accordance with ISO 4406
Systems with extremely high dirt sensitivity and very high availability requirements	≤ 16/12/9
Systems with high dirt sensitivity and high availability requirements, such as servo valve technology	≤ 18/13/10
Systems with proportional valves and pressures > 160 bar	≤ 18/14/11
Vane pumps, piston pumps, piston engines	≤ 19/16/13
Modern industrial hydraulic systems, directional valves, pressure valves	≤ 20/16/13
Industrial hydraulic systems with large tolerances and low dirt sensitivity	≤ 21/17/14

Pumps	ISO Ratings
Fixed Gear Pump	19/17/15
Fixed Vane Pump	19/17/14
Fixed Piston Pump	18/16/14
Variable Vane Pump	18/16/14
Variable Piston Pump	17/15/13
Valves	
Directional (solenoid)	20/18/15
Pressure (modulating)	19/17/14
Flow Controls (standard)	19/17/14
Check Valves	20/18/15
Cartridge Valves	20/18/15
Load-sensing Directional Valves	18/16/14
Proportional Pressure Controls	18/16/13
Proportional Cartridge Valves	18/16/13
Servo Valves	16/14/11*
Actuators	
Cylinders	20/18/15
Vane Motors	19/17/14
Axial Piston Motors	18/16/13
Gear Motors	20/18/15
Radial Piston Motors	19/17/15

Table 8.1- System-Based and Component-Based Cleanliness Requirements

6

6

	ISO Target Levels		
	Low/Medium Pressure Under 2000 psi (moderate conditions)	High Pressure 2000 to 2999 psi (low/medium with severe conditions')	Very High Pressure 3000 psi and over (high pressure with severe conditions')
Pumps			
Fixed Gear or Fixed Vane	20/18/15	19/17/14	18/16/13
Fixed Piston	19/17/14	18/16/13	17/15/12
Variable Vane	18/16/13	17/15/12	not applicable
Variable Piston	18/16/13	17/15/12	16/14/11
Valves			
Check Valve	20/18/15	20/18/15	19/17/14
Directional (solenoid)	20/18/15	19/17/14	18/16/13
Standard Flow Control	20/18/15	19/17/14	18/16/13
Cartridge Valve	19/17/14	18/16/13	17/15/12
Proportional Valve	18/16/13	17/15/12	16/14/11
Servo Valve	16/14/11	16/14/11	15/13/10
Actuators			
Cylinders, Vane Motors, Gear Motors	20/18/15	19/17/14	18/16/13
Piston Motors, Swash Plate Motors	19/17/14	18/16/13	17/15/12
Hydrostatic Drives	16/15/12	16/14/11	15/13/10
Test Stands	15/13/10	15/13/10	15/13/10
Bearings			
Journal Bearings	17/15/12	not applicable	not applicable
Industrial Gearboxes	17/15/12	not applicable	not applicable
Ball Bearings	15/13/10	not applicable	not applicable
Roller Bearings	16/14/11	not applicable	not applicable

Table 8.2- Pressure-Based Cleanliness Requirements (Courtesy of Hydac)

7

7

8.4- Examples of Filtration Solutions

More Info about Combined Return and Suction Filter review in textbook:

- **Figure 8.2 through 8.8**

8

8

Chapter 8 Reviews

None

Chapter 8 Assignment

Student Name: -- Student ID: ------------------

Date: -- Score: ------------------------

Assignment:

What is the adequate ISO cleanliness class for proportional valves?

Answers to Chapters Reviews

Chapter 1:

1	2	3	4	5	6	7	8	9	10

Chapter 2:

1	2	3	4	5	6	7	8	9	10
D	B	D	C	A	B	B	A	C	B

Chapter 3:

1	2	3	4	5	6	7	8	9	10
D	B	A	D	A	C				

Chapter 4:

1	2	3	4	5	6	7	8	9	10
D	C	B	D	A	D	C	A	D	C

Chapter 5:

1	2	3	4	5	6	7	8	9	10
B	A	C	D	C					

Chapter 6:

1	2	3	4	5	6	7	8	9	10
D	A	D	A	B					

Chapter 7:

1	2	3	4	5	6	7	8	9	10
D	C	A	B	C					

Chapter 8:

1	2	3	4	5	6	7	8	9	10

www.ingramcontent.com/pod-product-compliance
Lightning Source LLC
Chambersburg PA
CBHW052340210326
41597CB00037B/6207